GLIMPSES OF HEAVEN,
VISIONS OF HELL

'A comprehensive and absorbing tour of the political, economic, social, moral, psychological and religious aspects of virtual reality'

Nature Magazine

'A rattling good read . . . if you want to be better informed than your colleagues, read this book'

Information Exchange

'Carefully avoids becoming bogged down with technical jargon. Worth having on your reference shelf if you're into cyberspace issues and VR's history and probable future'

The Message

'A lively update, a wholesome desire to make our synthetic flesh creep'

London Review of Books

About the authors

BARRIE SHERMAN has already lived in different worlds having been a dental surgeon, an economist, a senior trade-union official and television and video presenter/producer. He has written innumerable articles and ten books, many analysing the impact of differing technologies on people and society. He was hooked on Virtual Reality six years ago when co-author Phil Judkins told him of American computer scientists suspending the laws of physics. He lives in north London. He now leads the European Commission panel of experts investigating applications of Virtual Reality.

PHIL JUDKINS has also occupied a variety of worlds since he graduated as a classical archaeologist from Cambridge. His interests in the integration of the needs of people, the work environment and modern technology have led to management roles in organisations as diverse as the Atomic Energy Authority, the Civil Service, local government, Rank Xerox and his present position in the financial services sector with Provincial Group. Among his interests he lists the history of radio and radar and supporting the development of the Royal Society of British Sculptors, in addition to a substantial lecturing and authorship programme. He lives in Wakefield, West Yorkshire.

Glimpses of Heaven, Visions of Hell

Virtual Reality and its Implications

Barrie Sherman &
Phil Judkins

CORONET BOOKS
Hodder and Stoughton

Copyright © 1992 Barrie Sherman and Phil Judkins

First published in Great Britain in 1992 by
Hodder and Stoughton Ltd

Coronet edition 1993

The right of Barrie Sherman and Phil Judkins to be identified as
the authors of this work has been asserted by them in accordance
with the Copyright, Designs and Patents Act 1988.

10 9 8 7 6 5 4 3 2 1

A CIP catalogue record for this title is available
from the British Library.

ISBN 0-340-60155-8

Printed and bound in Great Britain by
Cox and Wyman Ltd, Reading, Berks.
Photoset by Rowland Phototypesetting Ltd,
Bury St Edmunds, Suffolk.

Hodder and Stoughton Ltd
A Division of Hodder Headline PLC
47 Bedford Square
London WC1B 3DP

To Matthew, Nicholas and Clare –
the future is what they make of it.

Contents

Asterisks indicate updated chapters

Acknowledgments

As is customary we would like to thank those people who have helped to make this book possible, but in this instance it would mean having to thank virtually everyone we consulted, of whom we asked questions, and whose work we interrupted. So to avoid spending the next two hundred pages detailing a long list of names, and infuriating anyone we left out inadvertently, we would just like to say a loud 'THANK YOU' to everyone.

But there are a few people who just must be thanked by name, for without them there would have been no book. To Alun Lewis who dug, delved, discovered facts, suggested, argued and stimulated. To Bob Stone, and Andy Connell for their time, expertise, good humour and patience in the face of stupid questions, who, along with Charlie Grimsdale, allowed us to play with their latest equipment. To Bill Wiseman and the HITLab people for being so courteous to yet another visitor: to Bruce Dodworth for always returning calls, to Julia Sykes and Laura Page for keeping the two of us in touch, and not least to Kerry White of Rank Xerox who unscrambled a vital corrupted disc. To them all we owe a debt of gratitude. The advice and help was theirs, any mistakes or misjudgments are ours.

BDS and PJ

Part One

The Past and the Future

When the alchemist Ko Hung unveiled his newly invented gunpowder in Beijing, some seventeen hundred years ago, he was hailed as a saviour of the human race: he was using it to find the elixir of immortality. It was another five hundred years before it was used to make the world's first sophisticated fireworks, enlivening the human spirit with celebrations. But from that moment it took a mere sixty years for some bright, if malevolent, spark to realise its potential as a weapon. Its use had come round the full one hundred and eighty degrees from immortality to becoming an agent of untimely death – the bomb had been born.

Such radical and unforeseen twists of fate litter the history of technological change. Gutenberg's printing press was intended to promote goodness and worship through the wider readership of the Bible, not to incite the baser instincts through pornography or *Mein Kampf*. Strowger saw his first telephone exchange exclusively in terms of improving his undertaking business, not as one of the most potent instruments of social change ever. The Wright Brothers could not have imagined their Kitty Hawk machine developing into the Stealth bomber or becoming a main shrinking agent of the global village, any more than Herr Benz could envisage the accidents, deaths and misery, or the freedom of movement – if that is what a traffic jam can be called – that the automobile would bring. Indeed, Rutherford and other nuclear pioneers were dedicated to cheap and efficient energy, not Chernobyl, leukaemia, Hiroshima and the 'balance of terror'.

In short, at their outset, most technologies can be considered neutral. It is we, the people, who determine how, where and for what, they are used. And as the world grows more sophisticated, and its parts increasingly interrelated, so these decisions get more difficult and more important. Virtual Reality is the most recent of links in this long chain, and like these other fundamental changes – radio and television included – it will offer us visions of hell, as well as the more widely promised glimpses of heaven.

But today, approaching the twenty-first century, is not the best of times for science and technology. The view of hell is obscuring a sight of heaven. Holes in the ozone layer caused by chemicals, global warming, Chernobyl, Iraq's germ and warfare capability, and the general disgust at man-made pollution, have made the public wary. And another reality, scientifically produced, is a very powerful concept; indeed it could be said to challenge the unique position of God.

Virtual Reality (VR) is not only a powerful concept, it is also a powerful technology. We believe that it is the beginning of a very far-reaching set of technological and social changes. VR will not only affect computing itself, but an amazingly wide range of basic activities, including: education, training, medicine, healing, politics, travel, shopping, marketing, advertising, religion, design, war, communications, entertainment, psychology, philosophy, space research, industrial processes, employment, and leisure. In short it will have an impact on a considerable slice of our lives.

Many of the changes will be for the better, some will not. And even those that are conceived with the best of motives in mind will be shown to have their darker sides, while other excellent ideas will founder for lack of sufficient support. Virtual Reality will have gender repercussions – it works in a way more likely to appeal to women than men. It will distort existing power and influence channels. It will.

The Past and the Future

affect the way families and friends interact, it could even affect sex, it might well attract organised crime, and will almost certainly be exploited by the ruthless. In short, like electricity, antibiotics, flight, television and computing itself, it has the capacity to transform society, both industrialised and developing.

For a new technology in its very early stages, VR has received considerable media attention. While its name may be unfamiliar to some people, the image of computer games players wearing helmets more commonly associated with science fiction films is remembered by nearly everyone. Whether people like it or hate it, overall, they recognise that Virtual Reality is something special. Even so august a person as Akio Morita the Chairman of Sony has claimed 'Virtual Reality is the next big thing after Camcorders'. Yet for all the talk and hype, articles and programmes, no one has yet combined an analysis of what it is with what it can do; and what and who it will affect – the upsides and downsides, as it were. This book is intended to start the process of doing just that. We believe it is important that this should be done before we begin to make our mistakes, not afterwards.

We have written the book with the technologically inexpert, yet interested, firmly in mind. To the best of our ability we have kept concepts simple, jargon to a minimum and tried to convey our own sense of excitement, awe and foreboding. We hope you find it enjoyable, and if it stimulates you both to find out more, and keep an eye on what is going on, so much the better.

Virtual Introduction

When this book was published in 1992 we anticipated there would be so many changes in so short a time that it would be partly out of date the moment it was written, and we were not wrong. Over the past twelve months the original Virtual Reality company VPL has all but disappeared, yet smaller ones have flourished. Applications have come to market, some products have been improved, others

15

dropped, and research has been put on a sounder footing. Much of this is due to the fact that a growing number of large corporations now see in Virtual Reality a technology capable of solving problems they had previously considered to be intractable. In turn this has stimulated technical improvements in VR systems, and tentative moves towards a set of universally acceptable standards.

But among all the changes three important non-technical events stand out. The American presidential election propelled Senator Al Gore into the Vice-Presidency. While some may argue – with a modicum of truth – that this represents a diminution in his power, it certainly enhances his prestige, and his ability to influence the national political and social agenda. This is important because while chairing the US Senate Committee enquiring into Virtual Reality in May 1991, the then Senator Gore said, 'Virtual Reality promises to revolutionise the way we use computers,' and 'clearly the Japanese are serious about not only catching up but dominating this new field [Virtual Reality], just like the old famous VCR story. But this is going to be a lot bigger in its implications than was the VCR.' In other words Virtual Reality now has a powerful friend at the highest level of government in the country with the most expertise in the technology. And this has been demonstrated by the President's Office of Science and Technology approach to VR. It recently assembled a high-powered team representing 15 government agencies to evaluate the potential of VR and its importance and value to the USA. It issued a quick first report in May 1993 which stated, amongst other things ... 'VR is too important a technology to the nation to continue its unstructured research and development.'

The second event was that Europe woke up to the fact that yet another new technology was about to pass it by. One of the wide-awake directorates of the European Commission decided to find out whether Virtual Reality was capable of being used practically in socially desirable sectors. Although partly a technological investigation, it put the horse firmly where it belongs – in front of the cart. Instead of looking for a use for clever technical gismos, the EC is basing its study on possible applications of VR, so putting user demand first. As this is precisely how economics tell us products are developed, it is a very welcome approach.

At a different level, but none the less a watershed, SEGA announced that by year end 1993 it would be selling a keenly priced home VR game system; and if anything will drive the acceptability of VR it is the entertainment market. And at almost the opposite pole there has been an upsurge of academic interest in VR. A consensus is developing among scientists that the technology should be referred to as 'Virtual Environments' rather than what they consider to be the philosophically confusing Virtual Reality. But whatever its name the technology is being investigated more fully. Arguments are starting, some bitter and less than academic, and Professors of Virtual Reality are being appointed. Yet these may be but straws in the wind, to be blown away by a winter of market and technological failures. On the other hand they may be bringing us that much nearer the start of a new phase in human society, where for the first time we can slip into and out of the real world and into another as we choose – or as our employers dictate. Only time will tell which will happen.

We have added some of the more important developments to this updated edition in the form of Virtual Chapters added to the end of the relevant chapters. We have also completely updated and revised the technical appendix to make it more useful to those who may wish to dabble in their own virtual worlds.

1

Introducing a New Technology

Virtual Reality is not the easiest concept to grasp; there is more than a little truth in the suggestion that you can only understand it by experiencing it. Indeed, if this book had included photographs or line drawings, they would have had to be either holograms or the children's 'pop-up' variety. Trying to illustrate this three-dimensional, highly intensive, interactive, all-inclusive technology with two-dimensional illustrations would have been like trying to play Beethoven's Fifth Symphony on a comb and paper.

There is no doubt that Virtual Reality (VR) exists. It is neither the figment of a science fiction writer's imagination, nor the raving of a brain-fried, sixties drop-out. Outlined baldly, it sounds technical and boring; but then a violin concerto doesn't sound too exciting when reduced to the essentials of scraping bits of catgut with bits of horsehair. In both instances the magic lies in the execution.

Virtual Reality allows you to explore a computer-generated world by actually being in it. In other words computers are to Virtual Reality what the spinal column is to a vertebrate. Instead of looking at a screen you are enclosed in a three-dimensional graphic universe where you can affect what happens to the virtual world, just as you do in the real one. You can walk around, pick things up and put them elsewhere, bend, turn, climb stairs, play musical instruments or whatever, but with the added complication that the laws of physics can be suspended. It is the most dramatic, and potentially most far-reaching, computer development since the silicon chip itself. It is science fiction come true – it is science fact.

Virtual Reality is usually based on a graphic world, although film and video can be part of it. It has three components: it is inclusive, it is interactive, and it all happens in real time. That is to say you become part of that world, you can change it, and the changes occur as you make them.

There are three basic forms of Virtual Reality. The first, and best known, uses small TV screens and earphones in a Darth Vader (or Gulf pilot) helmet, and a glove (or a joystick, wand, or six-dimensional mouse). The second form is where video cameras place and track the image of a user (or users) in a virtual graphic world, in which they interact with virtual objects. A variant on this is to convert the video image into graphics and place this graphic image of the user in the virtual world. The third type is to take three-dimensional modelling, but either view it through 3-D glasses, play it on a flat screen (like a CAD package), or a large, curved or angled screen to get the inclusive, or what are known as 'immersion' effects.

Surprisingly, Virtual Reality is not new. The idea, and the component parts, are tried and tested. The clever bit was putting it all together in the first instance; and it is this combination that is still in its infancy. The head-mounted displays and gloves, both with sensors locating them in space, are bulky, clumsy and constraining. Graphics are still some way from being realistic or sophisticated, and time lags between moving the head or hand and the effect these have on the virtual world are still too long. Existing optics do not give a good enough peripheral vision, and sound could be made to dovetail into the world far more accurately. Furthermore, there is only a rudimentary sense of touch, and the boundaries of virtual objects are only now being constructed so that hands cannot pass through them.

Nevertheless, each of these deficiencies is being overcome. Research into every technical aspect of Virtual Reality is being carried out worldwide, in universities, especially in

America and Japan, in technical institutions, and in companies both large and small. Not only are researchers trying to improve existing techniques, in many instances they are developing entirely new lines of thought. For example, TV screen spectacles replacing helmets will be a step in the right direction, but retinal imaging (projecting a virtual image directly on to the eye) is a leap of staggering proportions. And it is not a matter of will we get an accurate sense of touch in virtual worlds, the question is when? Funding comes from the military, space programmes, governments (mainly the Japanese), convinced education or health departments, and from companies and venture capitalists. Science fiction writers recognised the value of Virtual Reality some time ago, but now trade and academic journals are starting to take it seriously, technical books are being written, TV programmes have been screened and not unnaturally it is featured increasingly on computer bulletin boards.

Despite current technical limitations, the first products are coming on to the market. Public and private demonstrations of practical applications have been successfully managed. Arcade games have been given much publicity, as has the possibility of virtual sex; the games have arrived, the sex has not – yet. The technology is starting to play a part in medical and manual skill training, and is a proven trainer for some military skills. It is being piloted in schools, is creeping into computer interfaces, has helped disabled people, is changing the preconceptions of the entertainment industry, has made the war machine more deadly, and is on the way to becoming the architect's, designer's and engineer's friend.

Other technologies are combining, and increasingly will combine, with Virtual Reality. Artificial intelligence (AI), itself just emerging from a long 'winter', neural networks, and nano, genetic and molecular engineering, are likely candidates to create synergy with VR. Not only will these connections enhance the abilities of both arms of the technology link, but the combinations will almost certainly spawn new products and services. But above all, High

Definition Television (HDTV) and fibre-optic cabling will meld with Virtual Reality, bringing its benefits, and its darker sides, directly into homes and businesses. And as both are known quantities, just awaiting political patronage or commercial guarantees, they will not be long delayed.

Even as this is being written the technology is changing, improving and diversifying. Laboratories and companies in the USA, Europe and Japan are refining VR techniques and equipment at truly staggering speed, matched only by the increase in the number of its possible applications. Much of the Japanese effort is concentrated on Virtual Reality becoming a major telecommunication device. The Advanced Robotics Research Centre (ARRC), based at Britain's University of Salford, is developing working telepresence robots at commercial prices. A VR advice and demonstration centre has been opened in Stuttgart, Germany. The large computer companies are exploring this, the first natural interface. The University of North Carolina (UNC) is refining its molecular manipulator and the Human Interface Technology Lab (HIT-Lab) at the University of Washington, Seattle, and the Massachusetts Institute of Technology (MIT), are carrying on their wide-ranging investigations into new products and concepts. The military is refining existing techniques, and developing new ones, and space programmes are waking up to VR's cost-saving potential. Smaller companies and institutes are approaching the worlds of art, medicine, space and entertainment, anxious to develop and exploit their ideas for this new technology. New and practical design tools are announced as frequently as new art forums. It is a small world in a state of rapid growth, and almost permanent agitation and flux. Today's theory, literally, is tomorrow's reality.

But the larger world is not so certain, eyeing Virtual Reality with a mixture of anticipation and fear. Most industries, or consultants to industries, are sufficiently intrigued to be trying to find out how they can use it – the huge numbers of mainstream companies attending VR conferences and briefings make that quite clear. Some have identified training as an important area, others are considering marketing and advertising, safety and disaster simulation devices, design, communications, enjoyment and health and healing. Each has its advocates, and each its detractors. Meanwhile two other sets of people are contemplating VR with a sense of anticipation. Virtual sex is not that far away. Pornographers have a new medium on which to play their limited repertoire, and 'New Agers' are welcoming Virtual Reality in much the same way that the Incas embraced Cortez, with unabashed enthusiasm – but hopefully not with similar results.

Yet there exists a residual unease, as if people sense that VR is not quite right. The feeling seems to be that one reality at a time is enough to be getting on with. To make matters worse, the reality everyone knows is unstable; even our atlases are having to be changed, seemingly on a daily basis, as nation conflicts with nation and the old Soviet empire splits asunder. But to a great extent the unease has been created by misrepresentation. The message most of the media has been putting across of games and sex, drugs and games, positions VR as a threat. And this latest threat comes on top of other, more real threats. Crack and narcotics, AIDS, the ozone layer, global warming and terrorism are making the world a hostile, even frightening place. At the same time a barrage of media coverage is telling us that smoking, drinking, eating meat, wearing leather, driving cars, or having cream on your muesli is antisocial, or is even courting death. Virtual Reality is another mixed message. At one level it is frightening, at another enjoyable. But it appears that enjoyable things kill you, maim you or destroy your relationships. Added to this it is new, invented by scientists whom you find it difficult to understand, and is endorsed by people whose excesses of the sixties are said to be so harmful.

This concentration by the media on the sex, drugs and rock and roll aspects of VR is tantamount to highlighting a rumour that Saddam Hussein keeps ten mistresses; all very headline-worthy, but ignoring the real problems that the man creates. How often have you heard it suggested that Virtual Reality can create concentrations of power? Or that its unique intensity could make it a dangerous weapon in marketing, especially of political and religious propaganda? Not too often we suspect; virtual sex takes precedence. Why is the subject of addiction among vulnerable people ignored? Or the ethics of virtual world use for treating psychoses? And who will use VR? Who will control it? Who will design it? Who will provide it – and why? And more than anything else, who will pay for it?

Funding is important; it implies control of the development and use of the technology. Technologies, of themselves, have few consequences. It is what we use them for, or, to be more accurate, what we allow them to be used for that matters. In Parts One and Two you will discover what Virtual Reality is, what is being planned, and who is doing what. Not that this will be straightforward. As the noted American scientist William Bricken has observed, 'Virtual Reality is now in the unique position of being commercially available before being academically understood.' Just as gunpowder changed the world, but not in the way Ko Hung intended, so Virtual Reality may not be confined to its original purposes. Not only did gunpowder not confer immortality, it improved our capacity to kill and wage war; it made conquest and enslavement easier and had profound effects on power structures. It also had secondary effects, changing jobs, industries, clothing and architecture, as well as destroying peace of mind. VR has the same potential.

Although it sounds like the subject of some sort of medieval disputation or a metaphysical concept, VR is very practical. It transforms the computer from its traditional role as a processor of data (numbers and words) into a machine which generates a different, visual reality. And reality is the crux of this technique. Some VR enthusiasts have argued that cinema and TV, paintings and theatre are all virtual realities. We shall not play sterile word games with rigid definitions – they lead nowhere and enlighten no one. But we shall lay down some ground rules, some yardstick against which we will be able to recognise and measure a Virtual Reality system should we step into it.

The user of a Virtual Reality world must be able to interact with it. Put simply this means that within the world a user must be capable of movement, capable of changing objects or things, and capable of dictating what happens next. And these must happen with almost no time lag between the command or instruction and the resulting action; in computer terms this is known as doing things in 'real time'. In addition, VR is a very intense, active experience. It is impossible to let it wash over you in the same way that a television programme or film can be watched passively.

The other key factor is that users find the virtual world believable. Myron Kreuger, one of the ancients of this fledgling industry – he is old enough to remember John Kennedy at first hand – has what he calls 'the duck test'. If someone ducks away from a virtual stone aimed at their head, even if they know the stone is not real, it shows the world is believable. This is known as 'immersion', and the aim of many VR practitioners is to achieve total immersion. Although not strictly necessary, this implies that a virtual world must be three-dimensional.

Because almost all its current component parts are well tried and tested, VR is insulated against fundamental unforeseen snags, making it an unlikely candidate to succumb to the Theory of Technological Insufficiency. This states that with new technologies we always overestimate what we can do, underestimate the time it will take, and fail to recognise the problems. Nuclear power is cited as the classic example.

The basic computer system, known quaintly as a 'reality engine', needs to be very powerful to handle the myriad computations necessary to provide believable graphics in real time. And when you enter this graphic virtual world it is through your eyes and ears, in a helmet. This has two small TV screens, one for each eye, set minutely out of synch, so as to give that everyday feeling of the world moving as your head moves. With associated stereo sound, a wearer can turn through a full circle, seeing, hearing and interacting with the entire surrounding world. In this sense Virtual Reality only exists when the user is in sensory deprivation from everyday reality.

We make computers do things by using controls (input devices), and the mechanism by which we 'connect' or communicate with computers is called an 'interface'. Electronic computers have been with us for less than fifty years, yet in that period they have been reinvented several times over; indeed they could be said to be in a constant state of flux. The first system of communicating with computers used removable plugs. This gave way to paper tape, then punch cards, which in turn were superseded by keyboards and then windows and mice in a two-dimensional office representation. Virtual Reality will form the basis for the next generation of computer interfaces, when the physical attributes (voice, hands, gaze or body) of users themselves will be the instruments of communication, all in a three-dimensional environment.

The system works by sensing where the user's head, hands or body are in space; the computer then tracks, and responds, to subsequent movements. A magnetic sensing device with a sensor in the helmet allows the computer to track the position of the head. So if you duck down the computer knows it has to produce a new set of graphics to give you a low-angle view, under a table, or, if you prefer, up a drain-pipe. If you turn your head to the left the computer knows it has to generate graphics to show the view at that point. Another favoured input device is the Data-Glove, with the magnetic sensor and fibre-optic cabling (developed by VPL in California). The computer is programmed to respond to the movements and position of the hand, which can pick up or move virtual objects. In addition, by making the appropriate hand sign the user can move through the world. For example, in one system pointing the forefinger will move the person forward; point two fingers upward and you open a door, and so on. Each system develops its own set of signals.

VPL have taken this idea a step further and developed a

whole body garment (the DataSuit, like a wired catsuit) which also transmits instructions to the computer – in this instance the user's entire body is the interface! And, proving the point about fast-moving technology, yet another Californian, Greg Panos, has developed 'body jewellery' to do the same job, before the DataSuit has even been properly marketed. (Because these changes are so rapid, current technical details and specifications – operational or proposed – are detailed in the Appendix.)

This is the first time that computers have been designed to respond to ordinary human behaviour. Up until now, humans have been made to think and behave like computers. Even technophobes and self-confessed computer illiterates will be able to operate VR systems with minimal instruction. And VR's high intuitive content will make it the first computer technique (other than text processing – really typing) which is as likely to be used by women as by men, if not more so. There are, however, other ways of telling a VR computer what you want it to do. Over the next few chapters we shall see that we can talk to it, write to it, look at part of it, or manipulate six-dimensional mice, wands, joy-sticks, three-dimensional space-balls or hand-grips like the top of a sports-car gear shift. All involve natural, everyday habits and movements. The office or workplace feel is decidedly absent.

Although they carry the same potential, the other two main strands of Virtual Reality are less well known, because they are neither sexy nor photogenic. (The *Face* and other high style magazines have been attempting to use VR equipment in photographic sessions.) The video camera system is flexible. It might show only hands on a desk, the user's head and shoulders, or the entire person. The user interacts with objects in this virtual world. At the work end of the spectrum these may be virtual pens, keyboards or spreadsheets; at the pleasure end a user can juggle virtual balls, dance and play virtual drums and xylophones. In terms of commercial use what is even more important is that in both this and the helmeted variety, a virtual world can be shared by other users – it can be a meeting place, a market place, a sporting venue, or even a battleground.

The two-dimensional screen variant is known as desktop VR. It is however proper Virtual Reality in that the images are generated in response to commands in real time (they are not just pulled from the memory). Incidentally, this can be done on commercially available PCs. Although the system is designed for an ordinary office, when the graphics are played on a wide-angle curved screen, with the user seated centrally, a very high degree of immersion is achieved, so it could have other uses.

It is also necessary to explain what Virtual Reality is not. It is not yet a fully fledged separate industry, although the dragon's teeth have been sown, and new companies are emerging every day. Nor, in itself, is it yet a clearly defined, standardised product. It is not just the combination of computer technology, video, sound tape and text known in the trade as multimedia. It is not the Back to the Future ride at Universal Studios, no matter how persuasive and brilliant that may be. It is not even Smellorama or the Feelies. None of these actually allows users to get into different realities, where their decisions and actions are capable of altering the realities themselves. The realistic rides are unchangeable, and while multimedia is alterable (to a point) it is demonstrably part of everyday reality. If anything VR is closest to a pilot's flight simulator, or its predecessor the Linktrainer, both of which have an intensity and immediacy which enable users entirely to suspend disbelief.

This intensity, along with its versatility and ease of use, will make VR a potent technology. In its own way it has the capacity to affect society as deeply as the telephone or television. Both have changed the way we communicate with our family and friends, changed how we view the world around us, and changed the way we approach business, politics, religion and personal relationships. VR has already given us arcade games and sophisticated armaments, training modules and surgery aides, but these represent only the topmost tip of what will be an enormous iceberg, perhaps even Antarctica. From marketing to engineering and from air traffic control to mind control, VR will play an increasingly important role in public and private life as we move into the twenty-first century.

Such a major change trails behind it a series of problems – perhaps better described as challenges. These will be in the moral, ethical, political and legal areas, emerging either singly or more probably in some combination. Never before have we been confronted with the *practical* consequences of inhabiting a different reality, and these consequences will be compounded when VR combines with other technologies, such as artificial intelligence or high definition television.

Even today, some states in America consider artificial intelligence programs to be legal entities in their own right. William Bricken is on record as saying, 'Virtual Reality programs themselves may have rights as they become more independent of human control.' And Kreuger points out a potentially dangerous mismatch: 'The machine [computer] is evolving faster than at any time in history – but people are not evolving at all.'

That these are glimpses of heaven is certain, but we must remember that VR was used first to deliver missiles more accurately, so that they would destroy and kill more efficiently. Although, subsequently, it has been looked upon as a saviour of lives, both physically and spiritually, military use still continues. Which will be the dominant force? Today, with the benefit of hindsight, gunpowder, or indeed nuclear power, would have been better left undiscovered. Will we think this of Virtual Reality in the next century?

But now, in the very early days, people have taken Virtual Reality to their hearts, particularly in America. It is, perhaps, a proxy for the American dream – to be at the centre, the President, a star in your own Hollywood movie. And that is what you can be. You can invent a world with yourself as its focus, invent a personality to go with it, and a storyline that makes the most of your new status. If it sounds too good to be true, that indeed may be the case. As history forever demonstrates, where the future beckons we follow, at times too uncritically for our own good. Hopefully this book will help us to tread the path with our eyes fully opened to the possible potholes and pratfalls.

So we shall crystal-gaze to some extent. Using what we have learned of the technology, and its potential from current research, we shall look at Virtual Reality over the longer term: its uses, changes, impacts, consequences and possibilities. In theory it can do so much. We can foresee its uses in medicine and schools, but what about being virtually imprisoned, or taking part in the Crucifixion; electronic LSD or virtual interrogations? Overall, will it be a force for good or evil – and who will decide? Who will benefit from it, and at what cost, if costs there be? What safeguards will we need, if any? Can the existing legal system cope with this new medium? Will we need to exercise moral and ethical judgments over its use? And are we just going to glimpse heaven before getting first-hand visions of hell – or will both exist only in our virtual worlds?

2

How We Arrived At Where We Are...

The earliest electronic computers were developed to break codes, calculate the trajectory of gun-shells, and perform other tedious, time-consuming but essential sums. These mid 1940s and 1950s machines were number crunchers. And it was for this reason they were given the name 'computers', a title once reserved for the people who were employed to make such calculations. For the first twenty years of their existence, computers lived down to their name. They were used, almost exclusively, to process and manipulate data, most of which was numerical. Indeed, outside the more rarefied university research areas, text processing did not become readily available until the late 1970s.

Although computer-generated graphics had become commercially available slightly earlier, in the mid 1970s, this was seen within the computer industry as an extension of the mathematics of engineering. However, in retrospect, computer-aided design (CAD) was a watershed. Not only did it revolutionise pre-manufacturing processes, it also hinted at the long-dormant creative potential of computers. Eyes were at last opened. Thinkers and doers, cartoonists and artists, printers and film-makers all realised that under their very noses had been a machine capable of mind-boggling physical representations. And once the cost of these machines started to fall, in proportion to the increase in their processing power, computer graphics really took off.

Ted Nelson, one of computing's truly original thinkers, has speculated on what would have happened had John von Neumann's name for computers been adopted. The godfather of modern computer architecture had wished to call them 'All purpose machines'. (Another computer pioneer Alan Turing also described the 'Universal machine'.) Nelson believes that had this captured the imagination, computers would have been regarded in a different light, and might have taken a totally different direction from the outset. Their primary use might well have been to create animated displays. But accountants and government statisticians got there first, and the graphic ability marked time until the 1980s.

A casual observer of the scientific and current affairs scene might be forgiven for believing that Virtual Reality is a technology of the 1990s. Certainly its media impact was negligible before 1989, and considerable from 1990 onwards. And, as if to reinforce this illusion, VR is typically *fin de siècle* – both outrageous and fashionable. It fits the nineties to a nicety. But it also has an element of the Beatles about it, of whimsy, of radical change, and of psychedelia. So it is no surprise to find that Virtual Reality was indeed a product of the 1960s, and like so much of that era it has taken twenty-five years to reach adolescence, let alone maturity.

The story of Virtual Reality reads like a modern fairy tale. Rather like the heroic, good, handsome princeling, there were many witnesses to VR's birth. But, after a promising beginning, the establishment disowned it, favouring its brothers AI and PC. There were long periods when it had to hide in basements or live rough on the streets, existing on scraps and waiting for its latent talents to be recognised. However, it was discovered and kidnapped by evil witches, masquerading as the military, and for many years was forced to do their bidding in secret dungeons. Aided by a friendly dragon, NASA, our hero escaped and, it being northern California, was then kissed into confident life by a friendly, dreadlocked courtier. Now, like the beanstalk, it lives happily ever after in the sunlight, acting as a beacon to those who believe in a radically different future.

Virtual Reality is an almost exclusively American story. Even the false starts – like Morton Heilig's early 1960s ill-fated, non-computer film-based Sensorama, which although it engaged all the senses (including smell) was never inter-active – were American. Despite the hype, there is no one piece of technology which can be called Virtual Reality; there are different strands. The computers themselves, head-mounted displays and optics, sensing, recepting and tracking equipment, sound and the sense of touch all come together. The trick was to visualise how this could be done.

Although the Philco corporation developed a remote stereo camera and head-mounted display as far back as 1958, the genesis of Virtual Reality is commonly accepted to be a paper, 'The Ultimate Display', written by Ivan Sutherland in 1965. In fact, Sutherland had laid a foundation stone before this. In 1962, he had developed Sketchpad, the first inter-active computer graphics system which not only produced images, but could also be used to command the computer. From 1967 and into the early 1970s, while at Harvard and then the University of Utah, Sutherland put his other ideas into practice, going on to create the first transparent, three-dimensional head-mounted display. But this was so heavy it had to be suspended from the ceiling, and showed only crude three-dimensional images of outline-style cubes. This mirrors the development of computers themselves: the first computer, ENIAC, was monstrously large, and many times less powerful and flexible than the feeblest of today's com-mercial laptops.

At around the same time Myron Kreuger was pursuing a similar, but crucially different, train of thought. While Sutherland, and indeed Nelson, concentrated on head-mounted displays, Kreuger approached the matter from a more psychological point of view. He delighted in surpris-ing people with the interactive, playful nature of the technol-ogy, using floor pressure pads, and getting people to draw on screens by using the *image* of a finger. But this was a period of multiple distractions. Vietnam and the politics of protest were in full swing – Kreuger remembers pushing a computer on a trolley through a full-scale campus demonstration! 'Flower power' was decaying into the fruiting stage, and some seeds were germinating into the Manson-style drop-out culture. The media had more than enough sensations to be getting on with, and had no need to bother with an esoteric computer development. Complementing this, and the public indifference to most matters scientific, a whole generation of young people had come to believe this type of research work was 'naff'. The technology behind Virtual Reality appeared to go into hibernation.

But there was another reason for the false start. Like so many pioneers before them, Sutherland, Kreuger and Nelson were ahead of their time. Their systems could not be made commercially with the technology of the day. Either reliable equipment had not yet been developed (miniature cathode ray, or liquid crystal display television screens) or it was too bulky or too expensive (computing power). Over the years, reductions in component costs have been at least as important to VR development as have breakthroughs in technology.

Virtual Reality is not the first, and nor will it be the last, technological whizz which apparently stalled, only to be kick-started back into life at a later date. Food irradiation was written up in the 1950s but only produced thirty years on, while robotics, neural networks and genetic engineering also have had lengthy gestation periods. Kreuger likens Virtual Reality research to the treatment of a good wine: 'It was kept in the dark for a long time.' The conservatism of academic publishing did not help. Virtual Reality was not considered to be a 'recognised' scientific subject. No journals specialised in it. Yet to get the subject widely known, and widely debated, it had to be published in the appropriate magazines. Either these did not exist, or the editorial boards had never heard of VR – and clearly did not intend to start. It was a Catch 22. Furthermore, academic life, especially in America, revolves around the axiom 'publish or die'. So why do VR research, and commit suicide?

Myron Kreuger kept the flame burning through the 1970s and 1980s, continuing his academic research. He also had the confidence to take his creations around county fairs and museums, giving participative demonstrations everywhere he stopped. But the head-mounted display technique marked time. At least it did as far as the non-aficionados were concerned. Behind the scenes, however, work continued on the component parts. Researchers in different parts of America were imagining the unimaginable. At the University of North Carolina, Professor Fred Brooks was building his team to work, among other things, on the tactile manipulation of virtually real molecules. The never-ending galaxy of stars at MIT, especially its Media Lab, went on their dizzy way, investigating gaze and other interfaces, force feedback and, of course, graphics. The Aspen map, a non-immersive, interactive experience allowing users to self-navigate through the town, was created in this quiet period. And there was always the US Air Force, with Tom Furness beavering away secretly on cockpit instrument displays. In hindsight it was a very creative period.

Then in 1985 the public were exposed to two seminal VR events. The California-based NASA/Ames Aerospace Human Factors Research Division told the world it had developed a new head-mounted three-dimensional display and Kreuger's Videoplace opened to the public in the Connecticut Museum of Natural History. The story behind the helmet epitomises the 'Elastoplast and safety-pin' aspect of early Virtual Reality research. Michael McGreevey of NASA/Ames needed a helmet-mounted display for a project he was about to start. Three years earlier, and with the minimum of publicity, the US Air Force had developed its own helmet for 'head-up' display research, and he asked how much one would cost. The million dollar price-tag set him to thinking, then improvising. He removed the liquid crystal display 'screens' (LCD) from two cheap Radio Shack mini-TVs, coupled them with wide-angle optics in a motor-cycle helmet, installed his magnetic tracking device, and connected it to image-generating and control computers. The world's first affordable Virtual Reality helmet – and system – had been made. It was not only cheap, it was relatively lightweight and portable – although, in truth, the picture quality reflected its price!

Videoplace was the public proving ground for the video camera variety of Virtual Reality. Visitors to the facility interacted with each other. They could fingerpaint, tickle each other, even combine in gymnastics, dancing, or juggling – despite the fact they were in different rooms. It was a huge success with the participants. The concept of the shared virtual space was now shared with the public, although it is difficult to grasp that both these developments were part of the same basic technology. They could not have been more different in design, or intent.

But Virtual Reality was coming together elsewhere. Away from the public gaze, and at about the same time, a glove invented a few years earlier by Tom Zimmerman so he could play 'air-guitar' (miming with no physical guitar, as if part of a karaoke act, but actually getting notes and chords) was developed jointly with Jaron Lanier, whose main claim to fame at that point was as an innovative game programmer on Atari's Moondust. It was not the first gesture feedback glove. Dan Sandine at the University of Illinois had designed one far earlier; and another, on different lines, had already been patented.

Modern Virtual Reality was born when NASA/Ames, in the shape of Scott Fisher, commissioned Lanier's company, VPL, to make a glove for their VR system. It was a good deal for both parties. For a relatively small amount of money it opened up NASA's potential to develop its own VR system and remote space robotics, while it gave VPL the chance to be taken seriously (although this was not its first sale). In technological terms it was a clever advance, combining fibre optics and the magnetic sensor. But in media terms it was a quantum leap – the public now had something sensational to hold on to. Magazines and journals, from the *Wall Street*

Journal and *Time* to *Mondo 2000*, started to pay attention to this new technology, mixing a serious technological approach with considerable amounts of hype.

But the mere fact that the basic VR kit was now assembled, and the public were starting to hear of it, did not mean that the technology was 'fit for human consumption' as it were. The lag between command and action was too great, the graphics too blurred and grainy. But as is the way with such technologies, these difficulties did not prove insuperable; techniques needed to be refined, rather than rethought. In the intervening years since 1986 there has been remarkable progress in providing better quality vision and response, with Lanier's VPL in the forefront.

The last few decades of the twentieth century have not been overly kind to scientists and inventors. Where once the Edisons, Marconis, Einsteins, Flemings, even Whittles of this world were lauded and recognised, their names household words, the media now prefer to concentrate on corporate financiers, movie stars and politicians. Jaron Lanier is perhaps an exception to this rule. This is probably because he looks anything but a scientist, with shambling dreadlocks and Oxfam trousers. Perhaps it is also because he has such an unconventional background. Despite no formal training he is clearly a formidable computer thinker, with a vivid philosophical and, more to the point, practical imagination. Perhaps it is because he and his company have set the pace in the business sector, providing the overwhelming majority of the world's VR hardware and software and developing new products as easily as other scientists write academic papers. But, whatever the reason, this musician manqué has become the media's choice as front man for Virtual Reality, and is seen often on the front pages of prestigious mainstream magazines.

But not all publicity surrounding VR's development has been favourable. When Timothy Leary, a resurrected 1960s bogey man, gave his blessing to VR, mainstream America and Europe shuddered. They had seen this before – the typical self-indulgent, hedonistic, Californian drop-out scene. Virtual Reality then attracted the attentions of the self-styled 'cyberpunks' and 'hackers', and suspicions were confirmed. It was sex, drugs and rock and roll all over again, only worse. This time it was dressed up in difficult-to-understand, electronic jargon. Lanier's appearance did not help matters. He just did not look enough like an ideal son-in-law. These reservations and fears remain, and indeed, as we shall argue later, some of them may well be justified. Nevertheless, VR has gripped the imagination of the young, and in terms of its future development, that is what matters.

As if his technical achievements were not enough, Lanier gave Virtual Reality its name. It could have been called 'Artificial Reality'; Kreuger refers to it this way. The Japanese call it 'Intimate Presence' and no doubt the French will coin a suitably Gallic *nom*. Lanier might have called it 'Synthetic Reality', or 'Synthetic Worlds'; 'Imaginary Worlds', 'Conceptual Reality', even 'Cyberreality'. If all else failed he could have used 'Cyberspace', the brilliant fictional idea of novelist William Gibson which, nevertheless, is bandied about in association with VR as though it actually exists. But despite, or perhaps because of the fact that Virtual Reality is not easy to comprehend at first acquaintance, its name is perfect. It is neither uncritically functional nor tackily quasi-scientific. Rather, it is poetic, mysterious, elusive – and what other type of name could describe another reality?

But it is ironic that after all the American work and research it was an Englishman working from Leicester in the East Midlands who unveiled the world's first mass-oriented, helmeted VR product. Jonathan Waldern, an electronics buff working with CAD/CAM, had lighted upon the technology in 1981. In his words, 'Computer-aided design was less design than documentation,' and while researching a better communication channel with computers, he discovered a paper on holographs and projections by Jim Clarke, one of Sutherland's students in Utah. It was the end of CAD and the beginning of VR for Waldern.

He designed his first system in 1984, and published the details (jointly with E. Edmonds) in the Pergamon Infotech

1985 State of the Art for CAD/CAM. The response was patchy;
at the time he was alone in this work on the eastern side
of the Atlantic. By 1988, in a garage, he and some associates
had developed the first VR arcade game prototypes.
(Although true, this reinforces VR's romantic image. Inventors work in garages, artists struggle in garrets.) Not wishing
to rush out an unreliable product he waited until the morning of Friday, 22 March 1991 to launch the world's first
commercially available leisure industry VR system – the 1000
SD – at Wembley in north London.

While all the noise was being made in America, quietly,
with neither fuss nor hype, Waldern's company, W Industries, had designed and produced its own hardware and
software for a fully immersive, interactive, helmeted, arcade
game. Then, within months, on 31 July the company placed
the world's first shared space, Virtual Reality game – Dactyl
Nightmare – where two players stalked and shot at each
other, in London's Covent Garden's Rock Garden club. One
week later Total Destruction, a game played by four drivers
racing each other around the same virtually real racetrack,
created the world's first immersive Simulation Centre at the
Trocadero, also in London. Photos of the helmeted players
were in all the newspapers. Virtual Reality had arrived, with
a vengeance.

A year later in spring 1992 W Industries appears to have
stolen a march on the competition. It has the current state
of the art in arcade games. But Waldern knows that what is
state of the art today is tomorrow's obsolescence. Neither
he, nor any other manufacturer, designer or researcher
involved in Virtual Reality can rest on his or her laurels for
more than an instant. With the rapid improvement in
graphics, optics, computers, sensors and displays, and with
wealthy corporations sniffing around the edges, who knows
what tomorrow's trade press might bring?

3

... And Where We Are Now

Today, Virtual Reality has reached public consciousness, and to a considerable extent has grabbed its imagination: so much so that in February 1992 the *Daily Mirror* could offer a W Industries VR game system as a competition prize. We know that in America it has reached the political arena: in May 1991 it was the subject of a Senate committee hearing. We know research is widening to cover new areas, and deepening to make systems more reliable, robust and acceptable. We know that although the pioneering, smaller companies are still involved, research is increasingly being funded by large corporations and enterprises – some, in Japan particularly, with active government encouragement. The 1990s are well on their way to becoming the decade of marketable Virtual Reality.

As systems fall in price, high street stores are being built around VR machines, ad agencies are using them in promotions, arcade games are catching on, VR systems are in schools and colleges, and design, marketing and walk-through systems are being used by the public. In America, doctors are practising bronchoscopy and other techniques, using virtual patients; in Germany design teams are using VR to 'feel' what their dreams will look like; in Japan you can design your ideal kitchen, and in England an estate agent will 'walk' you through your new, but as yet unbought, house. Telepresence robots, where a remote operator sees through the robot's 'eyes', have been ordered; virtual space stations have been delivered. Town planners, engineers, nuclear installations and hospitals have, or want to have them; defence departments and space agencies need them. Virtual Reality systems have become only too real.

Describing where we are in technological terms is more difficult. There is no such thing as an average, or typical, Virtual Reality system, any more than there is an average dog. But thankfully, although there are hundreds of dog breeds, there are only three basic variants on the VR theme: helmeted, video-based and flat screen. However, within this trio, different uses dictate different equipment. The components used in arcade games are nothing like those used in surgeon training systems. And to complicate matters further, there are times when any two of the three variants can be combined, as in remote VR-operated (telepresence) robotics, where the helmeted and video types merge. To overcome this diversity we shall outline a skeletal Virtual Reality system, and fill in the details as we come to them, as if we were doing a painting by numbers.

The bare bones of a VR system consist of one or more computers, sensors (or input devices), display arrangements and software. The main computer may be able to generate three-dimensional graphics or it may need to combine with graphic boards (or rendering engines). Users see these 3-D graphic images on the display. These may be augmented by sound cues, such as a door shutting, running water or a motor starting – and touch feedback through the glove or joy-stick. By using various input devices (which act on the computer's sensors) users interact with the images. The sensors detect, among other things, the user's hand, head or body movements, voice and video pictures. Each time the sensor detects a change, the entire computer system (the reality engine) generates a new set of graphics, appropriate to that change. For a system to be recognisably 'real', perhaps a person moving around an office, the computer must be able to 'sample' the sensors about sixty times per second! This gives real-time changes (apparently instantaneous, but really with the shortest time lag possible).

The user sees, or perhaps hears or feels, the change on the displays – and then makes another decision about interacting. In real life these decisions will be continuous and very rapid.

The effect is exactly the same as in the real world where, if you are looking at a very large painting then take one step to the left, you see a slightly different picture. A small part of your field of vision on the right has disappeared, and a bit more of the left of the painting has come into view. This is precisely what the computer works out for you in a virtual world – it gives you the appropriate scene. In the real world, step forward towards a desk and you get closer to it; in the virtual world the computer makes the image of the desk that much larger (and removes the more lateral pieces of the wall behind it) giving the illusion of getting nearer. (Do this sixty to one hundred times a second and you move forward smoothly through the virtual world.) Virtually step back, and the virtual images get smaller. Virtually stand on a chair, or crouch, and the reality engine will generate new images of the new virtual angles and perspectives on the virtual world. If you pick up a virtual cup, start an engine, open a door or throw a hammer, the reality engine does the calculations and creates the virtual images and sounds to fit each, and every, movement.

The dog analogy also fits computers. Great Danes and Dachshunds have very different looks, temperaments and capabilities; yet both are dogs. In the same way there are computers and computers, ranging from the pocket aide-mémoire for the executive who likes to pose, to the superfast, supercool Crays, the most powerful commercially available computers in the world. However, it does not need much technical expertise to realise that producing a visually complex and detailed virtual world requires more computing power than that needed to create simple cubes. Equally, common sense suggests that the time lag between a sensor detecting an instruction and the resulting update of the graphical world is inversely proportional to the power of the computer. In other words, the nearer Virtual Reality gets to everyday reality, the more powerful the computer needs to be.

The real world is made of atoms and molecules; a virtual world is built with 'polygons', the basic building blocks of computer graphics. The more polygons that go into a world, the finer the picture, and the more realistic the world appears to be. But increasing the number of polygons delivered per second needs increasing amounts of computer power. It has been estimated that the real world represents between eighty and one hundred million polygons per second. Today's reality engine computers can, at best, deliver seven to ten thousand polygons per second. On the face of it, the scale of this problem is immense. However, human perception is very adaptable; animated cartoons using as little as five hundred polygons every second are widely accepted.

It follows that supporting a single complex graphic of an office, with lifelike shadows and lighting, textured carpets and desks, and the obligatory rubber plant (looking less like a green jelly than usual), requires considerably more computer power than that needed to support a simple cartoon-style graphic. But then put the graphic into a three-dimensional interactive context. Now the computer must be able to generate graphics covering every single angle of view, and every single juxtaposition of movable objects. Clearly, the more complex the graphic, the greater the number of possible combinations, the greater the number of equations to be solved, and so the greater the power required. Add to this visual, sound and tactile sensors, displays and feedbacks, all being processed in real time, and the power requirement can become massive.

Computer power generally is measured in the number of instructions the computer processes per second (in computerspeak, millions of instructions per second, or 'mips'). Although used formally in Virtual Reality worlds, this is not entirely appropriate. A VR system is not instructed in the conventional way. It uses its sensors as windows on to the real world, so to some extent can be said to be

43

auto-instructing. In any event each instruction, however it arrives in the computer, is converted into a series of mathematical equations – and it is solving these very quickly that consumes all the power.

Time lags are crucial. Response speed is of the essence. The old transatlantic telephone calls, with a one-second lag between talking and reception at the other end, made for annoying, surreal conversations. But that nuisance pales into insignificance compared with a similar time lag between a movement and the virtual response to the movement. This can be physically disconcerting to the point of sickness; indeed, lags of over three hundred and fifty milliseconds have proved to be unacceptable. Also they make a virtual world that much less believable. There are two reasons for the lag. The first, a phase lag affecting graphics, has only recently been recognised, yet even so appears to be on the verge of being overcome; the other lies in the power of the computer itself. As a rule, the greater the speed, the greater the number of instructions and computer calculations per second, and so the greater the amount of computer power needed.

There is a third power issue. Stripped to bare essentials, Virtual Reality systems have three components: imagery, interaction and behaviour. Of these, behaviour is the greediest in terms of computing power, as it needs to solve extensive sets of equations, over and again. So, in a virtual world where objects have 'lives' of their own – filing cabinets open, pencils write, kettles whistle and keyboards function – a quantum leap in power is needed.

Believable virtual worlds cannot yet be run on pocket calculators, and a Cray is still too expensive and impractical. Middling computers are used instead. VPL, the leading company, uses a Mac and two Silicon Graphics Iris machines (one for each eye) for their RB2 'off-the-shelf' system; NASA/Ames runs VIEW on a Hewlett Packard HP 9000; other systems run on 386 and 486s, IBM PCs and work stations, and SunSparcstations.

Companies specialising in Virtual Reality software, helmeted or otherwise, run their wares on the relatively low-cost 486 series, the type of PC found in a company managing director's office – but used mainly by his secretary. Among them are Sense8 and Vream of America and Dimension International whose Superscape, a walk-through programme designed for architects or planners, is produced in Aldermaston, Berkshire. Its founder and managing director, Ian Andrew, explains how he does it. 'We're software specialists,' he says. 'We started by creating computer games, and moved into Virtual Reality later, so we're used to writing clever software ... making the most of any computer platform. With our 3-D graphics on flat screens the computer doesn't have to support head-mounted displays and other things ... all the power goes into running the software.' He might have added it only has to respond to a space-ball and a mouse. Nevertheless, it produces a genuine Virtual Reality, creating its world in real time; and as we saw earlier, it can be transformed into an immersive system by using a wide, curved or angled screen. Another method is to use a modern equivalent of the old 'magic lantern'. Two powerful projectors, set so they overlap marginally, beam the virtual reality (or video) through liquid crystal display screens on to a large metallised screen. When viewed through polaroid glasses, the resulting images have proved to be realistically three-dimensional.

Alternatively, Superscape files can be transferred (ported in computerspeak) into a helmeted system. Bob Stone and his colleagues did this at the Advanced Robotics Research Centre as part of their telepresence robotic research. The reality engine they use, SuperVision, is one of four 'reality engine' computers dedicated solely to Virtual Reality; the others are a Silicon Graphics machine, the Expiality from W Industries, and Pixel Planes developed at the University of North Carolina. (Undoubtedly, by the time you are reading this, at least another three will be on the market.) Super-Vision solves the VR power problem by a technique (in computer terms an 'architecture') known as parallel processing. A conventional computer runs through a job, doing one job after another, like a Ford-style car mass-production line. Parallel processing works on the principle that a large set of tasks can be reduced to semi-autonomous, separate, individual tasks, all going on simultaneously – and the whole assembled later in the process, rather like the autonomous group work system used in the production of Volvo cars.

SuperVision is made by a small company called Division; Charlie Grimsdale, its managing director, an ardent advocate of parallelism, explains. 'The construction of virtual environments is inherently parallel. Different sensors have to be sampled and processed in parallel to provide increased sample rates and increased intelligence in control.' Translated, this means fast, accurate, virtual world creation and responses. And Grimsdale knows that a large part of the future of VR lies in rigorous scientific fields, such as real-time fluid dynamics, where great accuracy will be needed, and his reality engine will be expected to provide it.

Division is an unorthodox company. Set in the determinedly low-tech West Country market town of Chipping Sodbury, with a botanist for managing director and a frighteningly young staff, its technical eccentricity comes from its use of transputers to provide its parallelism. Transputers were invented at Inmos, a 1970s company started with British government support in an attempt to lift Britain into the ranks of the big players in the world chip market. While this did not come off, Inmos did invent the transputer, and its founder, Iain Barron, is now chairman of Division. Transputers are very powerful microprocessors (chips) that not only compute, they also communicate with other transputers or processors. Their great strength lies in the fact they can be assembled and used like Lego. The total workload is divided between various sensors, inputs and displays, and if it becomes too heavy for the system, additional blocks can be added. It gets around the obstacle that arises when non-parallel computers run up against the limits of their available power. In theory there are few limits to the power that transputers can bring to bear.

Another way of achieving parallel processing is to link

thousands of cheaper chips. UNC's Pixel Planes brings together two hundred and fifty thousand processors. This massive parallelism is dedicated to a visual display. UNC specialises in 'handling' virtual molecules, and fitting them together. This not only requires a good visual display, the operator also needs to 'feel' whether the fit is right. UNC have developed a tactile feedback system, and the reality engine has to provide the power for both accurate sharp graphics and a predictable sense of 'feel' – both in real time. And this brings us to the second requirement for a Virtual Reality system, good displays.

The Pixel Planes display is huge, creating a feeling of immersion. (When we sit in front of a screen which is filling over sixty per cent of our field of vision, we feel as though we are in the scene – not just viewing it.) Smaller, concave or wide-angled screens can have the same effect. An ordinary cathode ray tube monitor, like a TV set, is sufficient for two-dimensional viewing of three-dimensional graphics; and nowadays it is possible to have the use of columnar or shuttered glasses to convert these graphics into three-dimensional images. Every cathode ray tube picture, on your TV set for example, is made up of individual points of light, known as pixels. Pixel Planes has a quarter of a million of them – each with its own microprocessor!

Current helmeted display technology is based on colour LCDs, the type you get in Watchman TVs or hand-held computer games. Each eye has its own small screen, containing around ninety thousand pixels, enough to give a reasonably sharp image. The user views the screens through wide-angle optics, with shutter, lenticular or holographic lenses, filling the user's total area of vision. There is a simple, but fundamental idea behind the lenses. They restrict the image to one eye at a time. When done sufficiently quickly this brings a three-dimensional quality, but if it is too slow the image flickers, and when immersed in an intense virtual world, this can be most uncomfortable, even dangerous.

When your car turns around a corner, the outside wheel has to travel further than the inside wheel, because they are set so far apart. The car solves the problem by employing a differential. In one of nature's stranger tricks, our eyes are set apart, so when you turn your head in real life the images nearest to you turn more than those further away. To reproduce this motion 'parallax', the LCDs are adjusted to show a very slightly different image to each eye at the same time.

Auditory displays have not been as high on the research agenda as visuals. Yet auditory cues are most important in life; listening for cars is a vital part of road safety awareness, and the Apple Mac sounds are vital to its success. Deaf people are handicapped profoundly. Current VR systems use stereo sound. You hear an engine starting, a door opening, or the sound of drumstick on kettle, but none of them has yet managed to pinpoint sound sources accurately, or their cadences satisfactorily. The technology exists, however, to produce infinitely better aural cue and sensing systems. Jonathan Waldern says he sat in a darkened room listening to the RCA 'binaural' sound system play the sounds of his hair being cut, and even heard the breath of the hairdresser snipping away with the scissors along the nape of his neck. Cues can be imaginative and valuable. For example, it is possible to let a busy pilot know a plane is running out of fuel by having the cockpit resound to a noise like the gurgle of an almost empty bath. (This is no new idea: tea-breaks at Xerox's Europarc are signalled by transmissions of rattling cups.) Aural displays also can be based on voice synthesis packages.

Both the sound and sight systems are mounted in a helmet. VPL produced the first commercial helmet called, imaginatively, the EyePhone. Since then at least four others have been marketed, some more expensive, some heavier, some better fitting and some better ventilated than others. But the helmet (more properly called a head-mounted display) has two other functions. It keeps out the 'real' world, both making the VR experience more vivid and aiding immersion. It also hosts an important additional component, without which this type of Virtual Reality simply would not work. It has a sensor which can track the position of the head in space.

It is all very well for you to look left, or upwards, but the reality engine computer has to be told; otherwise it cannot generate the correct set of graphics. A clever system, called the Polhemus, based on electromagnetic disturbances, was one of the original NASA/Ames breakthroughs in VR. Just as swallows follow electromagnetic lines when migrating, the system works by sensing a disturbance generated in the electromagnetic field of the helmet-mounted sensor, relative to a source of the field, which is placed within three feet or so of the user. The sensor transmits its information to the computer, which then adjusts the virtual world to the new position(s) of the helmet, and user's head. The Polhemus, however, has disadvantages. Long lag times, interference, and a very short range made the search for a successor inevitable. New systems are coming to market. One of them, rejoicing under the name of Flock of Birds, uses four receivers, takes one hundred samples every second within an eighteen-foot-diameter range, and promises to overcome these difficulties.

The counterbalanced stereoscopic viewer can also tell the reality engine where a user is looking. The user views through two eyepieces, like a pair of binoculars, with two small, cheap, cathode ray tube screens and wide-angle optics. It is head coupled, not head mounted. The binoculars are attached to a mechanical linkage like an anglepoise reading lamp that allows free movement, which microprocessors then transmit to the reality engine. Although it looks unwieldy, and only works within a very small radius, it is perfectly acceptable for desk use, and does allow easy access. In other words it is perfect for quick peeps, which is more than you can claim for the helmeted system!

Hand-eye co-ordination is so vital to real life that we use it without realising – and only discover its importance if it disappears. Even the simple things – eating, dressing or washing – are impossible without it. This is such a powerful metaphor for the real world that it was inevitable it should be transferred to virtual worlds. Although other gloves are being developed, the DataGlove, a hand sensor – again produced by VPL – was the first. It combines the Polhemus with fibre optics running along the length of each finger. The magnetic sensor tracks the hand in space; the fibre optics track finger flexing movements. The beauty of the glove was that it appeared to promote a natural way of doing things.

However, virtual life has proved to be more difficult to accommodate than real life. The glove has to serve two functions. By making different shapes and signs with the fingers, or gesturing, the user can move forward and back, up and down. But the user is also expected to use the glove to interact with objects in a natural way, picking up a cup or opening a cupboard. This combination of the 'surreal' and 'real' has proved difficult to master. What is more, this problem has been compounded by unreliability and unreality. The hand graphic is poor, and there is a general clumsiness, partly because you cannot feel objects – as yet. This lack of sensation feedback must be rather like suffering from leprosy. But much attention is now being paid to gloves and other devices which enable the user to have some sense of touch.

If you can do it with the hand, why not the whole body, ran the argument. VPL rose to the challenge, as always, and produced a full body DataSuit. This works in precisely the same way as the glove, and has the same advantages and disadvantages, but with the added problem of tangled cabling. Nevertheless, the idea that the entire body can act as a sensor for a computer system has mind-boggling possibilities. A sample of a puppet/master relationship using a full body suit, although not in a Virtual Reality context, can be seen in the Hollywood movie *FX2*.

Hollywood has always had a love affair with technology, especially computers which engage in lengthy conversations. HAL in *2001* is probably the most enduring, possibly because it exhibited human vices, not just virtues. And voice is, of course, the other major bodily-based VR sensor. You can command a VR system to do things. 'Open' may open doors, windows or files, depending on the scene and software. 'Start' may well start a motor. But what is different is that some commands will apply to the user. Say 'fly', and the user will fly; 'forward' will be the cue to move forward in the virtual world. It has a dream-like quality about it. The idea is not new but is very effective. Other sensor aids include treadmills, fixed bicycles and rowing machines, which enable users to walk, run, cycle or row – rather than float – through virtual environments.

Not all sensors have to be human-based; the video-camera is a perfect example. Interactive telepresence robotics may depend on video pictures, often converted into graphics. Powerful computer facilities also have to be provided for this operation. In other areas, medical training or surgical techniques for example, this need not happen – graphics may be superimposed on video pictures. Indeed, the Kreuger version of Virtual Reality requires a video-camera to position a user in much the same way as the helmeted variety uses Polhemus head and hand trackers. But sensors also can be non-visual. X-rays or infra-red, radioactivity or laser scanners, ultrasonics or radar are all used in telepresence robotics where visibility is zero – on the sea-bed or in a smoke-ridden environment.

With the glove still being evaluated and developed, movement and interactivity (grasping or moving objects) within virtual worlds has to be undertaken by other means. The 'six-dimensional' mouse is one of the favoured methods. The name refers to the six degrees of freedom available in Virtual Reality, as against only four in the constrained real world; and in truth the mice are really only three-dimensional. They work in precisely the same way as ordinary mice, except they navigate through a three-dimensional world, and with a click in the appropriate place, they lock on to objects. However, instead of moving files, as in the normal window operations, a user can then open doors, move desks, type letters, indeed do anything that the virtual objects have been programmed to do.

However, mice are not the only input agents available

today. Wands, hand-grips and space-balls are all capable of achieving the same effects. Instead of moving the mouse through space in the direction users wish to go, a series of buttons on the grip have to be pressed. One will take users forward, another back, one up, one down, and still another will lock on to objects. The space-ball, a ball-on-a-stand which can be rotated as well as moved in two dimensions so giving excellent mobility, has no buttons and has to be used in conjunction with an ordinary mouse to get locking. As things stand at the moment, the glove may appear glamorous, and it is certainly a crowd pleaser, but these other input devices tend to be more accurate and reliable. For those who, for one reason or another, are incapable of using these devices, a range of lip, swallow, blink, muscle and gaze centred inputs are in existence, or are in development stages.

Although it is a truism to say that a pair of scissors with only one blade is one of life's most useless implements, Virtual Reality hardware without appropriate software can rival it. Without software none of the hardware, however ingenious, will work. VR is a great consumer of programming time. Not only do complex operating and control systems have to be written, but the software for the virtual world itself is formidably lengthy, not to mention tedious to write. And as of now, some programs actually have to be written as if for two dimensions, and then transformed afterwards into three. This is equivalent to being given a videotape in the Betamax format. There are specific packages available. VPL offers Swivel, a colour solid modeller; Body Electric, a real-time animation product; and ISAAC, providing more realistic images. That there will be others, indeed specialist languages like MINDSHELL, goes without saying.

As things stand today there are two distinct ways of approaching Virtual Reality. A virtual world can be purchased 'off the peg', or can be created specially by software designers, either of which can be used for teaching, simulations and games. Alternatively you can create your own worlds. And it is certainly possible, even with today's technology, to network both approaches. VPL has crossed the Pacific; games with up to four players are in arcades; the Kreuger video version has been proved over several miles, while an education project has linked seven children into the same world. All that is needed is a local area network (Ethernet) or public telephone lines, preferably fibre optic-based. And as time and technology progress we shall have many more options.

This basic design distinction raises a series of fundamental points about the control of VR, and its impact on users and non-users alike. It is a subject we shall deal with in some considerable detail, for it may well have profound repercussions. But what is of interest here is that companies are starting to provide packages for do-it-yourself (DIY) virtual worlds. In America Sense8 produces a WorldToolKit which requires someone with previous CAD experience to manage. In Britain, Dimension International has produced a Three-Dimensional Toolkit selling for less than the London price of a very ordinary meal for two, which can be used by people who have never before laid eyes on a computer. If they possess basic hardware they can construct their own simple virtual world in a matter of minutes. Is this the future?

So, where are we today? The technology is still in its infancy – and there are damaging flaws. Graphics are not good enough. Time lags are still too long. Optics are not up to scratch. You can walk through walls. You can pick up a cup, and not feel a thing. The equipment can be as temperamental as the legendary diva, and is still generally far too expensive for everyday use. But each and every one of these drawbacks is the subject of intense research and work; the problems are being overcome. Virtual Reality is poised for its major breakthrough.

Virtual Chapter 3

Having pointed out that VR equipment is far from reliable, we must ask if it has improved over the past year. Well, it is still diva like. As one researcher says, 'If the hardware doesn't fall over then the software crashes ... but only when we're showing it to someone important.' Large research systems have so many different boxes, linked by losable cables, it is surprising so little goes wrong. Some equipment makers, however, are designing products with the user in the background, if not yet fully in the forefront of their minds, while the really unreliable equipment and gadgets are being unceremoniously discarded. But it has to be said that while the last year has seen an explosion of applications, VR equipment has been in a period of consolidation – with one exception.

As one would expect, computers, or reality engines, have improved markedly. This is a function of the fall in computer prices combined with their increasing power. (Over the past few years it has taken just fourteen months on average for computers to double their speed; at stable prices.) Indeed, Jonathan Waldern claims the true measure of computer power is 'mips per $'. Some of the specialist computers are awesomely powerful. Pixel Planes 5, the most recent of the University of North Carolina's giants, is to be replaced by PixelFlow, capable of delivering over ten million polygons per second. And Princeton's Sarnoff Research Centre has developed yet another massively parallel processor, the Princeton, soon to be upgraded and renamed the Sarnoff Engine.

But these are specialist one-offs. At the top end of the research and applications market Silicon Graphics has leap-frogged ahead. In quick succession it produced a Reality Engine, an updated Reality Engine II, and now the Onyx, a computer with more power, speed and graphics flexibility than any other in its price range. Other specialist computers from Division and W Industries have also been refined and given more speed. And as the PC market now uses the Intel 486 chip, at prices lower than last year's far less powerful 386, it provides workable platforms for home or business VR at reasonable cost. With the Pentium chip around the corner the entire range of VR platforms is, if anything, exceeding predictions with diminishing lag times and an ability to handle increasingly complex virtual worlds.

Not so visual displays. Despite a large number of new displays and viewers, much work remains to be done before they can be considered to be adequate. Various helmeted displays (known as head-mounted displays or HMDs) have come on to the market. Some are ingenious, with a flexible spine to fit the back of a user's head or a display unit that pivots out of the way without taking the helmet off. Others use cathode ray tubes with one thousand line displays, but tend to be heavy and suitable only for research work. Virtual Vision's TV Viewer has a single LCD display with a curved mirror, giving an image floating eight to fifteen feet in front of the user; it weighs only five ounces and overlays a virtual image on the real world. But this is heavy compared with Crystal Eyes ultra-lightweight LCD spectacles-style 3-D viewer, or the micro-HMSI which looks like black Mafia-style wraparound sunglasses, weighs under three ounces, and also can be used as a see-through device.

Gloves have proved to be expensive, cumbersome and highly fallible. However, where a new technology is concerned the trick is to learn from mistakes, and new gloves are being developed in the US, Italy and the UK. The Italian versions have force feedback, while the Vertex Cyberglove, designed and produced in Wales, threatens to undercut the price of the others by at least a factor of two, while improving reliability. The US now has a Greenleaf light glove with a battery-powered interface and cut-out palms and fingers. Magnetic tracking devices have also improved their range, and the lag time between movement and registering on the visual display. Indeed, one of them 'Flock of Birds' can now be used to act in a similar way as has been proposed for body jewellery, that is to track a human body without a full DataSuit. Ultrasonic trackers, video-based trackers and optic trackers are all making an impact, the latter the most likely to provide lag-free, extended-range tracking. Meanwhile, all manner of wands, space-trackers, mice and other interactive gadgets have been produced or improved.

The race to find a universal standard software for VR is

warming up. The present frontrunner is WorldToolKit from the US company Sense8. The most widely used software package in the world, it accounts for about half of all serious immersive VR systems (over three hundred and fifty) although it is completely outsold by Dimension's products, including Superscape. These, however, are used more for desktop or home/personal VR. Division is also in the race with its dVS2, and there is a move to get it recognised as the European standard, as is W Industries with its new software system. And in the pit-lane, waiting to see how the race unfolds are the HitLab with its VEOS, and AutoDesk.

Other VR software is also developing. Canon UK's RenderWare gives high-quality graphics, removing the direct dependance on ever faster, more powerful hardware. It works on a variety of PCs and work stations using parallel graphics 'pipelines'. Part and parcel of software packages are authoring tools, and here a very significant name crops up: IBM's VR-DECK. Whether this is just an aberrant gesture, or the start of a deeper VR involvement, only time will tell, but if the latter it will move VR on to a different plane.

4

Working for Tomorrow – Today

Bob Stone, who heads the research at the Advanced Robotics Research Centre, knows more than a thing or two about Virtual Reality. A prolific speaker at seminars, a writer, practical telepresence researcher, and solver of conceptual VR problems, he is probably more aware of what is happening in research, in systems, and more importantly what they might be used for, than any other person in Europe. So when he says, 'Virtual Reality is compelling, but poorly executed . . . a form of sensory deprivation . . .' and, 'Users have to rely on their psychological tolerance or persistence to accept, and enjoy, sparse visual and auditory stimuli . . .' we have to take notice. Clearly, those involved in VR recognise its flaws; they know quality could be better. That is one of the main reasons for the increasing number of research projects in so many countries around the world.

The genesis of a new idea or product is a strange phenomenon. Luck, or at least being in the right place at the right time, plays a large part in its success. One could say that Franklin was lucky to be flying his kite in a thunderstorm or Fleming fortunate to have contaminated his petri dish. But that would miss the point. Both had the genius to recognise the implications of what they saw. Today, money is also an important ingredient in success. More than a few brilliant ideas have been delayed, or indeed abandoned because of a lack of sufficient funds for basic mock-ups or development, although, conversely, experience suggests that throwing money at a problem has never guaranteed its success. But intuition, considerable lateral thinking, a deep understanding of what underpins the current technology, sheer hard work (as Edison pointed out), and creativity are also important factors. Designers working on the new Cray supercomputers are said to visualise their solutions, in much the same way that sculptors interact with their clay or stone. This creative or artistic bent is even more appropriate to Virtual Reality research because of the heavy involvement of visual and spatial elements, and the pioneers, Sutherland, Kreuger et al, displayed all these talents in abundance as they turned the received wisdom of computing on its head.

It is rare for large organisations, or their senior employees, to be directly involved in innovative research. Whether this is due to 'safe' recruitment policies, or because lateral thinkers generally do not seek such positions, is a subject which itself would repay research. And all too often such organisations also display an inbuilt inertia which militates against anything running counter to the corporate ethos, including new products and ideas. When JJ, the stereotype chairman in the British television series, *The Fall and Rise of Reginald Perrin* says, 'I didn't get where I am today by shilly-shallying about with new ideas,' he is reflecting the prevailing view of corporate man.

It follows that radical new ideas most often have to be generated and developed within small companies, on shoestring budgets. However, later developments, diversification into new applications and proper marketing are taken on by far larger, resource-rich organisations – an example is Henry Ford and car production. So when a technology is ready for take-off it needs a patron: a larger company, a venture capitalist or a government. Five years ago Virtual Reality was in this position. While it remained in the laboratory or exhibition halls, it was effectively side-lined, and when, for example, Myron Kreuger tried to break out, his ideas were far too removed from the experiences of engineers and computer people in mainstream organisations. To compound this, VR's first patron, the US Air Force, demanded, and got, anonymity.

But the moment NASA was involved, the whole idea gained industrial credibility. Almost overnight the world started to beat a path to the doors of those individuals, small companies and labs where the VR torch had been kept alight. It has now reached the point where some of these pioneers are spending more time at conferences and seminars than at their jobs. Meanwhile, the list of conference attendees reads like a copy of 'The Times 1000'. Ford and General Motors, Boeing and IBM, ICI and BAe, Bell and Anderson Consulting, Kajima and Nomura: they come from the engineering and commercial giants, the leisure conglomerates, and from advertising and marketing, to hear and see the small number of experts. As Charlie Grimsdale says, 'For every fifty articles and conferences there's only one firm doing something real.' And this interest is being maintained, despite the variable quality of Virtual Reality; clearly the potential is outweighing its current limitations.

Those attending these conferences have the history of technology as a good guide. Virtual Reality is merely the latest in a long line of mould-breaking products that were presented to the public, even though they were less than fully developed. Just compare early 'cat's-whisker' wireless with today's FM stereo; waxed cylinders with compact discs or the original television pictures with High Definition Television – and these are only in the communications world. If you look more widely – at the start of pharmaceuticals, aeroplanes, computers, even ball-point pens – there is a pattern of subsequent, dramatic improvement in performance, ergonomics, reliability, or almost whatever measure you care to choose, including *real* prices (adjusted for inflation).

Two interrelated mechanisms appear to be involved: the market, and research and development, with the emphasis on development. A market forms in the new product. With small production runs, and relatively crude equipment, there are few barriers to entry – companies proliferate. But those selling inferior models either fall by the wayside or are forced to invest to get improved quality. However, now the companies which were successful come under pressure from those which took the trouble to improve, and in turn they have to improve. They could cut prices, but if quality and reliability are important factors this will not do them much good in the longer run.

So quality ratchets up, and customer expectations ratchet up with it. The more successful companies merge or take over less successful ones. Seeing the profits, new players come into the market, bringing additional resources. They can afford proper development budgets. And development (rather than basic research) is what large companies do so well. The total number of companies falls. The technology becomes more user-friendly, as the handful of internationally based companies compete on customer service and quality in a market where prices are roughly the same over a range of different qualities. This happened with cars, telecoms, cameras, and with radio and TV receivers – and there is no reason why it should not happen with Virtual Reality.

There will be one major difference, however. Virtual Reality is an enabling technology. Certainly it will create games in arcades and the home; and yet, among other things it will probably underpin the relaxation market, but for the most part it will be used to make things happen with other technologies. It will combine with telecommunications, television, robotics, teaching, surgery, and aeronautics. VR's development will be bound intimately to the development of other technologies and techniques. Virtual Reality will be a combination of new consumer product, new computer interface, new component part, new service provider, and above all a new communications medium.

But before it can be properly combined with anything, the basic technology has to improve, and that is why today's research is important to tomorrow's products. It can be divided into two distinct parts. Research to improve the technology itself, and research into different areas of use. This chapter is concerned with the former.

Japan aside, most activity in this area is still confined to small companies and operations where individual skills count. It is true that large companies are buying expensive basic test hardware from VPL and Silicon Graphics in America and Division and W Industries in Britain; IBM, AT&T, Bell, BT, BAe, and GEC, are among them. But only large Japanese companies, and NASA, Boeing and the American military, are building their own test-beds and systems. When Senator Gore of Kentucky held his Senate hearing on Virtual Reality in May 1991, his main witnesses, all leaders in the field, were from specialist institutions. All three have featured in previous chapters: Jaron Lanier of VPL, Professor Fred Brooks of UNC and Dr Tom Furness, once of the US Air Force, now heading the Human Interface Technology Laboratory (HITLab) in Seattle. Government and NASA scientists were also in attendance, and it was one of these who estimated the total NASA research bill for Virtual Reality to be less than one million dollars. Given the billions of dollars the defence industries have spent on research, it has to be said that NASA has made spectacular returns on a paltry outlay.

All three institutions are engaged in fundamental research, concentrating on displays, sensors and software. But they are far from being alone. There are other centres in America, as well as Japan, Britain and Europe where equally important work is being done. Indeed, the USA has lost its monopoly position, and probably its overall lead as well.

The equipment, helmet, glove and body suit might be an essential part of the fun for games players, and very discriminating fetishists, but overall it has to be seen as a bar to further commercial progress. It has been argued that people are happy to wear special gear; after all motorcyclists, skateboarders and skiers use helmets and safety-wear; surgeons wear sterile clothing and nuclear workers dress like astronauts. But this ignores research findings that indicate most people use their workplace at least as much to socialise as to work – and helmeted VR puts a complete stop to that. Furthermore, if Virtual Reality teleconferencing is to succeed, and the Japanese NTT – the largest telecom company in the world – seems to believe it will, the eyes must be seen. This drawback is recognised widely, and much work is going into the display and sensor areas.

Visual displays are improving too. Both the Flight Helmet from Virtual Research and LEEP Systems' Cyberface 2 with its dual resolution mini-screens, are head and shoulders better than the original models. But the future may lie elsewhere. The person who holds the TV remote controller has power, and other viewers in the same room can get irritated. Sony realise this, and currently are developing TV screens on spectacles, so that people in the same room can watch different programmes. All it needs is a multiple switch device. And there is another factor. High Definition Television (HDTV) receivers will be relatively expensive. If, however, HDTV is introduced on spectacles, costs could be cut dramatically, and this technology would then be available cheaply to drive Virtual Reality use.

HDTV development is important to the future of VR in another way. Not only does it give far sharper picture quality, it is based on an entirely different concept of today's television. Its digital base will make it as valuable to future telecommunications as to television and multimedia producers, although its main impact on VR is that it can support computer technology. This means that it will be able to incorporate a computer, or interact directly with one. HDTV is in a state of confusion at present. Japan and America have competing systems, whilst Europe has two of its own. The struggle for the dominant standard is joined, reminiscent of the Philips, Betamax, VHS video-format battle of the 1980s; but whoever wins, by the mid 1990s Virtual Reality will have gained.

The HITLab, on the edge of the University of Washington campus in Seattle, is in the forefront of Virtual Reality research, with projects ranging from biomedicine, education and aeronautics to art. Fortunate enough to have engaged the practical and financial interest of the local large manufacturer, Boeing, it has also put together a consortium of companies, most but not all American, to fund further work. At present it is pursuing two separate lines of visual research. It has developed a tiny cathode ray tube half an inch in diameter, with two thousand scan lines, above HDTV standards. Unlike the Sony system this is intended for use in high-grade scientific work.

The other line of research reads like pure science fiction. Dr Tom Furness is working on 'retinal imaging'. This would put a picture directly on to the retina from a pinpoint source, a retinal scanner, using low-density lasers. This is not a matter of pixels; it is a matter of eye physiology, of rods and cones. There would be no screen of any description, no helmet, just spectacles, perhaps without the sort of glass we know today. And, by varying focal lengths, it should be possible to have mixed virtual and real images, at one and the same time. Despite the privately expressed doubts of US government officials and some of his rivals, Furness is bullish about the prospects of a breakthrough in the near future – and by this he means well before the end of this century.

Stephen Beck is also confident. If Furness's research seems like a twenty-second-century technology, then Beck's appears to be centred far into the twenty-fifth. His invention, Virtual Light, produces optical sensations directly on the eye using electrons, not photons. Beck believes we can get much higher 'bandwidth' (more information at a time) by channelling information directly to the brain. His Phosphotron works with goggles, not implants, although he foresees a future of designer-style spectacles. Beck claims highly successful tests, although only for repeatable light flashes, not pictures as yet. Even so, should this be validated, there will be two lines of research driving us towards screenless displays. And as there will also be two lines of research leading to screens in spectacles – or contact lenses – it is evident that the days of the bulky, intrusive, head-mounted display helmet are numbered.

The second technological research area concerns position and body sensors. Position sensors are crude, but becoming less so. Early experiments with sonic sensors, abandoned because of unforeseen echoes, are being restarted in both Japan and America, while the Polhemus is giving way to more sophisticated devices, also based on magnetic fields. The basic glove is being refined, and the fibre-optic attachments being made more robust. Sensors based on optics are being developed at the HITLab, and there is also 'body jewellery'. This is worn in the usual places, ankles, wrists, fingers, neck, ears, waist, forearms, chest and hair, with radio linking. Although the principle is precisely that of the DataSuit, it is certainly less cumbersome, less constraining, and possibly more accurate for certain functions, especially in show business.

Body jewellery could represent the best concept yet for full body sensing, aside from using video images for stereographic conversion, *à la* Kreuger. Judging from the research effort being deployed to overcome practical problems associated with converting video images to high-grade graphics, especially of faces, large Japanese companies are convinced of the potential of the video method. Their Ministry of International Trade and Industry (MITI), a superministry equivalent of a combined British Treasury and Department of Trade and Industry, with a smidgeon of the Department of Employment and Foreign Office thrown in, has even started a 'Telexistence Lab'.

Work on other sensors is carrying on apace, much of it for the defence industry. In the mid 1960s it was realised that pilots in modern warplanes were subject to what was euphemistically described as 'visual information overload'; translated this meant too many dials to cope with. To solve the problem, which has worsened over time, a 'virtual cockpit' was designed for the US Air Force. It makes use of the pilots' senses of hearing and touch, as well as sight, and uses symbolic rather than just analogue displays. Among other things it pioneered 'head-up' (on the windscreen) displays, along with voice and gaze sensors. Research into

this area continues in America in the Supercockpit programme, and is replicated in Britain at various sites of British
Aerospace, and in companies like Rediffusion and GEC. It
has not been entirely successful as yet. Today's VR systems
may be accurate enough to pilot a virtual spaceship to a
virtual planet, but this falls short of the detail needed to fly
a real modern bomber on a mission. It is, however, this
type of research – where standards need to be of the highest
– that will drive Virtual Reality technology on to a higher
plane.

Gaze and voice sensors are not new, and neither are they
confined to virtual systems. Gaze sensors work by bouncing
narrow beams of infra-red light on to eye pupils, and noting
the reflected angle. This tells the computer precisely where the
user is looking, selecting items on a menu for example.
However, if voice sensing is also available a user should be
capable of navigating in a virtual world just by looking and
talking. You gaze at where you want to go then give a direction, 'Forward', or 'Up', leaving the hands completely free
to perform other tasks. However, gaze and helmet displays
do not fit together as easily as egg and bacon, and considerable work is being done at BAe and other places to overcome this. And voice is a problem, although of a different
magnitude. Most people tend to swallow the ends of their
words, making it difficult for the computer to understand.
Sir John Gielgud's Shakespearian delivery would be the one
of choice, but as this is not yet a requirement for qualification as a military pilot, restricted vocabularies are sub-
stituted.

Sounds make worlds real, although not always pleasantly;
noisy neighbours playing loud music are a prime cause of
local arguments and assaults. Yet music also can 'soothe
the savage breast', reaching out to the violent, disturbed or
autistic when all else fails. And try to imagine a movie without its incidental music or its squeaky door, creaking
floorboard sound effects. It would lose a lot of its impact.
Sound not only augments vision, it can take over from it
when the visual senses are overloaded; it is even possible

to listen to a movie, and still know what is happening. Sounds, noises and hearing are so obviously important to our everyday reality, they will have to be improved in Virtual Reality.

Although we hear sounds three-dimensionally, our different ear shapes mean we all hear something slightly different. In addition, we normalise very loud or soft sounds. So programming the location and sources of sound to suit all users is difficult. The HITLab is researching ways of getting sound to react to people, rather than the other way around, while also using virtual reflective surfaces to give echoes, making location that much easier. And it is building up a library of sounds with three categories of information: where a sound starts, stops, and where it comes from. Using this it hopes to add computer-generated sounds to its worlds, more realistically and usefully than before. The Convolvotron computer system, developed by NASA/Ames and Crystal River Engineering, uses the same tracking devices as helmets to source its binaural sounds, the type that convinced Jon Waldern he was having his hair cut! Sound technology is coming together, in Milan, Italy where Professor Gardin is digitising virtual voices, and at other sites in America, Britain and France, although it is still expensive in terms both of money and computing power.

One of the more irritating VR drawbacks is the lack of sensation when handling virtual objects. It detracts hugely from believability, and impairs the work of engineers, designers and medical personnel, among others. In Salford, where the ARRC has formed a consortium of industrial companies, albeit on a more modest scale than at the HITLab, Bob Stone is co-ordinating work on a force feedback device, the Teletact II Glove, with Jim Hennequen of Airmuscle Ltd in Cranfield and VPL in California. The glove uses thirty compressed air pockets and micro-capillary tubes which, when inflated, push against the fingers and palm of the hand, giving the impression of holding or touching an object – which can be as slim as a pencil. These combine with the fibre optics and position sensor of the DataGlove in a double

Lycra layer. It is not yet foolproof. A human error caused a compressed air pocket on a prototype ARRC force feedback, six-dimensional mouse called the Commander to explode very noisily in Stone's hand during a demonstration – scaring the audience of one far more than Bob Stone!

The real trick will be to combine the force feedback sensation with a Virtual Reality experience. Researchers at ARRC are working on mapping the boundaries of virtual objects (their outside edges) with their appropriate sensations. For example, the reality engine will be programmed to reproduce the sensation of picking up a bottle when the boundary of the graphic bottle is touched by the glove. In turn the computer will send instructions to the compressor to fill the air sacs of the Teletact glove in the right strengths and areas to give the wearer that physical sensation. Work is proceeding at ARRC on building up a library of sensations.

Similar projects are being pursued in Turin, Italy with tactile gloves; Rutgers University with piston, micro-actuator gloves; Grenoble in France with force feedback virtual musical instruments; and in Japan for scientific manipulations and telepresence. A different route which makes use of exoskeleton equipment (complex engineered systems of levers and sensors) has been developed at both UNC and by Margaret Minsky at MIT's Media Lab. They are getting surprisingly delicate, accurate feedback from heavy, bulky equipment. But neither appear to be developed sufficiently to get the reductions in complexity, size and price that would enable them to enter mass markets, although their importance as test-beds as well as for specialist medical use and one-off or laboratory work cannot be overstated. There is, however, another way to get a virtual feeling of touch.

Tactile simulation sounds a bit tacky, the sort of thing advertised on stickers in London's Paddington Station phone booths, next to 'Minor alterations to gents' clothing' and a telephone number. In fact, as we shall see, it might become this, but meanwhile it is a feedback device in its formative stage. One method is to use crystals which vibrate when a current is passed through them (piezoelectric cells);

another uses electromechanical forces. An American company TiNi is a pioneer of this research (it made a tactile glove for the US Air Force back in 1985 using special alloys to create fingertip feelings). More recently TiNi created arrays of small blunt pins pushing against the skin, with a force and frequency dictated by a computer. The button on a Simgraphics three-dimensional mouse has been equipped with nine of these arrays, which will soon feature in a glove.

The device is called a 'tactor', and this illustrates a minor, but diverting point. Virtual Reality is adding a technological Runyonesque strand to the English language. Eyephone, tactor, Cyberspace, teledildonics (virtual sex – back to tactile simulation), virtuphone, speaktacles (sound spectacles), Simbin (VR system), automagically, telepresence, and even goggle-roving have all featured in articles or speeches. And to cap them all, while addressing a VR conference, the author Howard Rheingold referred to TV coverage of the Gulf War as 'disinfotainment'.

Realistic touch/feel sensations, visual images and sounds will be the touchstones of Virtual Reality, determining its ultimate success. Users must be convinced they are in real situations – even when they know they are not – and this boils down to having sufficiently powerful reality engines. Today's supercomputers operate at over one billion floating point operations (flops) per second, but at best an ordinary personal computer can manage just one hundred thousand. However, on average, computer speed has increased one thousandfold every ten years, and experts are confident this will continue. So hardware is not seen as a bottleneck, although its cost is. Supercomputers, such as the Cray, can solve many of VR's power problems, but their initial, and subsequent, running costs are prohibitively expensive. Not long ago the laser was described as a brilliant technology waiting for a use. The transputer was in a similar position until Virtual Reality. Now it provides VR's relatively low-cost, flexible, computer power. Research into this and other parallel processing techniques continues in America and Japan, as well as Chipping Sodbury.

However, research into better displays, sensors and reality engines only addresses part of the problem. Graphics must look good, and behave in a believable, consistent manner. This is less a matter of research than applying different techniques. Fractal geometry is, and increasingly will be, important in making natural worlds look natural. Plants, mountains and fields all benefit from this approach, which softens the ubiquitous polygon. Texturing overlays the graphic surface with pictures of the real thing – a painting, a clock-face or the fuzziness of a carpet – and this technique makes the world of artefacts look more natural. Off-the-shelf packages to texture virtual surfaces are available now, but to become realistic, new programming tools will need to be devised. It is one thing making a floor, tiles or a machine look natural, but it is quite another making soft, pliable objects like a dishcloth look like a dishcloth!

Nothing is more disconcerting than inadvertently flying through a wall, or seeing your hand disappear into a ball or a spanner. Other than for special 'magic' or experimental worlds, objects will have to be programmed to display their real physical characteristics. Rubber balls will have to be made to bounce more than wooden ones, steel must 'give' more than iron, and water must have a different specific gravity to oil. And there is a real difficulty with text and numerals: neither transmit well in VR. While this may be alleviated by improved visual displays, giving sharper edges, less flicker and better resolution, in all probability new typefaces to fit the medium will have to be produced.

The difference between a good painting and a masterpiece is often traceable to the way the master has used light. And, just as good lighting can make an indifferent movie into a special one, bad lighting can ruin the believability of a virtual world. New software has been developed by Thorn EMI to provide real-time lighting for virtual worlds. Objects are programmed to 'remember' the lighting effect associated with them using a technique known as 'progressive radiosity'. The effect of this is to incorporate a real-time, virtual approximation of movie industry lighting practices.

Indeed, ad agencies have come to realise that to make their confectionery products look really toothsome, Superdelicious Smarties, they must use graphics and graphic lighting, not the real thing. They have discovered that graphics can be more real than reality. French and American companies are also developing virtual light systems, which is just as well, as without them virtual worlds will remain crude cartoons.

For most business purposes, virtual worlds will have to be believable. Yet we all know that flying like Batman is, sad to say, not the everyday way of getting about London or New York. Yet we are expected to tolerate flying in VR. Although treadmills and cycles can be used with appropriate sensors to simulate walking or riding, they are not comfortable, convenient or practical ways of moving about in virtual space. Furthermore, it can take fifteen minutes to teach a person how to 'fly', and this will be an unacceptably high price to pay in terms of people per hour usage on public demonstration machines. Placing video-cameras on users' heads has been tried, but has both weight and range limitations. Dimension International has developed a walking representation, but this requires the individual to be represented by the walker. In other words users would be seeing a version of themselves, unless telepresence techniques (seeing through the eyes of the virtual walker) were used. Even though this involves a double dose of virtuality it might be less disconcerting than trying to fly. Perhaps pressure pads, or a body jewellery anklet combined with a Polhemus will do the trick.

Flying is only one potentially disorienting factor. We have very little idea what being immersed in Virtual Reality will do in terms of physical or psychological health and safety. Bob Stone at ARRC, along with Edinburgh University, is just embarking on research into the psychological toleration of exposure to VR systems. But to make the equipment ergonomically satisfactory, and to lay down guidelines about health and safety, far more research needs to be started – and quickly. For example, are there real re-entry problems?

Some people believe there are; at least one company, W Industries, has banned car use for a period after lengthy VR immersion.

Networks make the modern industrialised world tick; imagine life without telephones, television, and automatic cash dispensers. Virtual Reality promises to add to the scope of networks. As several people can inhabit the same virtual world at the same time, it can be used in much the same way as a videophone. However, because a large volume of information needs to be transmitted, it requires a large bandwidth, generally larger than the traditional telephone co-axial copper cables can provide. The Swedish Institute of Computer Science, based near Stockholm, is researching into broadband networks using fibre-optic cabling, with Virtual Reality very much in the forefront of its researchers' minds. Bellingham, a town in the state of Washington, has taken cabling a step further. US West, the regional telecom company, is fibre-cabling the whole town, one of the explicit purposes being to allow schools to use Virtual Reality systems. And as fibre optic is interactive it will also enable the sharing of city-wide information, adding to Virtual Reality's potential.

Holography is being investigated seriously in America as a possible virtual image representation system. American institutions are also researching sensing mechanisms based on tiny muscle movements. But it is work being done independent of Virtual Reality research which may well have the biggest impact on the VR systems of the future. Whether the most important links will be with artificial intelligence, fifth generation and bio-computing, neural networks, genetic engineering, nano-technology or perhaps branches of a yet undeveloped technology, only time will tell. Given the history of twentieth-century science, it is highly likely that synergy will develop from linkages of this type. But another, less physical area of research may well be the most important of all.

Virtual Reality is the first conceptual, almost intuitive, computer system – that is why it is believed that women will be able to understand and operate its systems more easily than men. It can impart information by analogy and metaphor. So why have lines of data when their meaning can be represented by a wheatfield or a mountain range? Why use text if sounds and colours can provide the same information? But we need to know what analogies will work, and on whom they will work. We hope you are enjoying the book, but have you asked yourself how you actually understand it? Indeed, how do we understand things, learn things? What are the mechanisms by which we associate sounds, colours, words, images or shapes with other, even intangible, things, like birth, death, marriage, happiness or sadness? What appeals to what senses – and why?

Without cognitive research we will not be able to use VR at anything like its full potential. In one respect this may be a good thing. In the hands of a Goebbels, a Stalin, an unscrupulous Ayatollah or Christian fundamentalist, who knows what might happen? But we would also lose its contribution to education, training, health and the creative arts. Many of the people working with Virtual Reality today realise that what is at stake is nothing less than its future. Although some work is being done, at the University of Aberdeen in Scotland – but mainly in America, it is nothing like enough.

So where are we now? NASA/Ames illustrates the position. It has a test-bed system called VIEW – Virtual Interface Environmental Workstation. It uses a head-mounted display, a glove, speech recognition and a three-dimensional sound system. All are state of the art, but VPL and Division, MIT, UNC, ARRC, HITLab, the Japanese and the defence companies are all trying to improve on what NASA has got. Sadly, official European agencies are committee-ridden, leading to hopeless technological compromises for 'political' reasons, and horribly slow response rates. The European Space Agency is only now commissioning research, while the European Commission research arm, Esprit, has just put together a multi-country consortium; although whether companies will now be prepared to divulge technical information to their rivals is open to question.

We have outlined the main lines of research for the very good reason that they represent the future of Virtual Reality. At the same time we have ignored the wilder shores of fantasy, such as brain implants. With the knowledge of what is being done we can make better judgments as to where, how and on whom Virtual Reality will have an impact. It is clear that it will look and sound much better, control will be more accurate and easier, the equipment will be far more sociable, and its costs will fall. We will be able to feel and touch, so it will be more believable. We will get it at home with fibre-optic cables, or cheap, mass-produced 'home reality engines', and it will be tailored to our needs. In other words it will be an altogether better, more commercial product. But for what?

In Part Two, we shall be looking at products and services available now, and considering some which we can expect in the foreseeable future. This will include research, especially the applied research in Japan, where they are throwing money at problems – and confidently expecting it to come back, with interest. But even though VR is a fledgling industry, it is an understatement to say that competition is virulent. Lawsuits are flying. Yet, outside the defence industries, research information was still being shared across the board until recently. Sadly, as is the nature of these things, this openness is disappearing. The scramble to optimise market share and profits is just starting.

Virtual Chapter 4

If anything, activity in fundamental VR research has increased in pace, depth and extent over the past year. While most of the same labs and companies are involved, new ones have been coming on the scene. For example, although the Meckler Directory of VR R & D organisations lists fifty-five establishments in Europe this is a massive understatement, omitting almost all the application and user-driven research and activities. Europe alone must have nearer a hundred and fifty working research efforts. In the UK the university computer department without a couple of researchers, perhaps graduate students, doing VR research with some form of VR kit is now very much the exception. At the same time, Taiwan, Singapore, and Australia are all starting research in VR, some, but by no means all of it, academic.

But the tendency to reinvent the wheel continues with, for example, very similar research on telepresence robotics being carried out in Germany and the UK. Indeed, work commissioned by the EC through ESPRIT not only duplicates work in individual European establishments, it even duplicates itself! But it cannot be faulted for imaginative acronyms. FIVE is the Framework for Immersive Virtual Environments, and others include GLAD-IN-ART (gloves), GRACE, and DESIRE. Genuine EC visionaries do exist. Michel Richonnier, a director in Directorate XIII, wants European scientists to build a virtual brain, if necessary in co-operation with American and Japanese VR researchers.

Despite having labelled it as a next-century technology, perhaps the most far-reaching of current research projects took an unexpectedly giant step forward when the Retinal Scanner made its public debut in a proof-of-concept demonstration in Seattle in autumn 1992. Although bulky in its present form, and with low resolution and a limited field of view, it successfully 'imprinted' both text and two graphic-based demonstrations. It may well turn out to be a *fin-de-siècle* technology. There is also a high-resolution 1000 × 1000 LCD which puts a display on glass and other material, not just silicon. Both have enormous potential to make VR more widely acceptable.

Although the Japanese and Europeans are engineering away in a myriad of labs, there have been no reported fundamental breakthroughs as yet. However, there are some interesting lines of development. ARRC is experimenting with a laser pointer which interacts with a projected virtual world so that there can be interactivity at large demonstrations. Another far-reaching area of research is the University of North Carolina's optical tracking ceiling. It consists of ceiling tiles with infra-red displays embedded in them and a helmet with four cameras pointing at the ceiling. It can locate a user to within 2mm. ATR's Communications Systems Lab in Japan has come up with a method of digital puppetry (or animation) based on human bodies. It uses video shots of key points of a head and body and compares them (in real time) with the same points on a standard body. Potentially this could make cumbersome full-body suits, or even body jewellery, superfluous. Eye-tracking and force-feedback research continues, especially in the military and space areas. But researchers are beginning to realise the complexity of the combined sense of feel, touch and weight – it involves joint movements and the positions of limbs in space – and realise they must understand it before they can replicate it faithfully. Considerable amounts of money are now being diverted towards this end.

As it becomes apparent that VR is serious, there is an increasing interest in its effects on users. SRI International in the US is running a series of experiments, one of which showed that although people adjusted to frame rates as slow as one to four per second, rates between five and twelve per second cause discomfort (especially in the eleven to twelve region). Mel Slater of London University's Queen Mary and Westfield College (QMW) ran a series of immersive system experiments. If his discovery that users with a predisposition to travel sickness had a greater degree of 'presence' or immersion was surprising, then so was the clear gender difference. Women with a virtual body and a DataGlove felt more immersed than a control group (using a 6-D mouse and no virtual body), while conversely men in the control group were the ones who felt more immersed. However it was unsurprising when Slater found that loss of realism reduced the sense of 'presence'.

Work is being done in the US, Japan and Europe on what makes VR realistic. One group at the University of Edinburgh showed the need for fine textural detail, but their main research, which produced uncomfortable results about HMDs, has created dissension. It suggests that because we cannot focus and adjust our peripheral vision simultaneously – and objects are at a fixed distance in VR – inevitably there will be eye-strain lasting for some time after leaving the virtual world. (Research at SRI suggests a similar problem.) While there are loudly expressed doubts about both the research itself and the quality of their test HMD, no one disagrees with the Edinburgh conclusion that VR system builders should now 'take a user-centred approach'. It is needed. As Slater points out, 'Wearing an HMD means that you are legally blind, according to the American definition.' And because it now believes that VR may take off, the UK Health and Safety Executive has launched a study of the effects of VR on the physical and psychological health of users.

Part Two

What is on Offer Today ...
and the Day After

5

The VR Industry and Consumer Products

Virtual Reality is an industry on the edge of crossing the magic barrier that separates dreams from commercial reality, yet at the moment it is more reminiscent of the old Wild West in the gold-rush days. Desperate to stake their claims, the hopefuls are racing towards the horizon; elbowing, pushing, and not worrying overmuch if their wagon has rolled over a rival's foot, so long as they get there first! But this is the way enemies are made and feuds start.

Similar equipment is being marketed, and some people and companies believe it is too similar. With everything being so new, patents so blurred, experience so limited, expertise so rare, and information having been so readily shared, it is the litigation industry that is starting to thrive. This is a recent departure; a sense of humour used to prevail. When one of the pioneers, Eric Gullichson, now of Sense8, was running Autodesk, he tried to register Cyberspace as a trade mark. The author who dreamed it up, William Gibson, threatened to retaliate by registering the name Eric Gullichson. The attempt was dropped. Perhaps the current absence of this form of imaginative riposte means that money and cynicism are taking over, and that the magic barrier has, indeed, just been broken.

There is no doubt that the point has been reached when the free movement of information and readily available mutual help has become just a pleasant memory of more innocent times. Even university laboratories have formed companies to exploit their work, or have promised their funding consortium members first option on developments. There is a sad inevitability about such moves but, as happened in the American West, ultimately the pioneers have to stake out their positions. And will the analogy continue? Will the big boys use their muscle on the homesteaders? Will there be offers that cannot be refused, the present-day equivalent of damming the creek? Certainly the large corporations are standing on the sidelines, watching, waiting to see who and what develops, trying to spot potential winners. In February 1992 a pre-emptive defence organisation, the American support group Virtual Reality Homebrew Association, was created to prevent small company innovation 'being taken over by the giants'. It will have to work very hard to succeed.

Getting to know what is happening in the VR scene should not prove to be difficult for the giants; they can just dip into the developing network of information sources. Given the nature of Virtual Reality it would have been strange if computer bulletin boards and networks had not taken up the subject. The WELL, with its several thousand subscribers, runs its own Virtual Reality section and acts as a transmitter of Usenet; the largest subscription network service, CompuServe, now runs VR pages for non-professionals, while programmers can get VR ideas from BIX. On the old-fashioned print side *Cyber-Edge*, a bi-monthly Bay Area-based magazine, covers VR almost exclusively, as does *Virtual Reality Report*, published nine times a year by Meckler. Oddly, mainstream computer magazines and papers are scant sources of information, probably because VR is seen as a distraction from 'serious' business: national newspapers such as the *Independent* or *Sunday Times* are almost as knowledgeable. Finally there is *Mondo 2000*, a very Californian glossy which mixes eclectic science, music, art, sex, fashion and strange food advertisements into a wonderful newish-age cocktail. Not unnaturally it has featured VR heavily.

Are the giants glimpsing heaven through these publications, or is it up there hiding behind clouds of uncertainty? Well, although it is getting lighter, what we are seeing is anyone's guess; anyway what large companies call heaven, other people think is hell. However, by the end of 1991 it was clear that the public was seeing something different. There had been a breakthrough – Virtual Reality had invaded the high street. A screen-based game using Mattel's Power Glove (based on VPL's DataGlove) had been sold in America for the past year, while in December Mandala, a video-based VR entertainment system already on sale in North America, became available in Britain. At the same time, the first shops of a chain called Virtual Reality, owned by the retail group ERA, opened simultaneously in Leicester and Cardiff. With a futuristic decor built around a W Industries arcade game, on which customers are invited to play, the shops sell high-tech moving image electronic equipment, camcorders and the like, under the slogan, 'Technology to set your Imagination Free'. The stores will demonstrate VR systems as they become available, and, as they believe it is potentially a large market, intend to be identified with all VR and multi-media products. Although this is the first attempt at direct merchandising using VR, it is by no means the first marketing ploy.

Callscan, a telephone equipment company, successfully used VR as a marketing tool at a 1992 trade show in an attempt to get a concept, rather than a product, across. But VR works well with products. Advertising agencies have latched on to the public's fascination with Virtual Reality. It appears to stand for youthfulness, sex, the future and fantasy. It appeals to the young, to the fashion-conscious, to science fiction addicts and the adventurous. So far it has been used to advertise a Japanese truck, men's perfume, biscuits, sweets, and sports clothing – and with its appeal, alcoholic drinks and cars cannot be far behind. Every time it is used in this way it extends the public awareness of the technology. But this exposure concentrates on arcade games, while the bulk of VR's probable uses are left unmentioned and untapped.

This tapping has already started, however. It has been suggested that Matsushita, the huge Japanese electrics company which also owns department stores and the American entertainment giant MCA – which in turn controls Universal Studios – once bought a construction company purely to give itself a captive outlet for its prodigious selection of kitchen and household electrical equipment. If this is true it is one of the most heroic examples of vertical integration ever conceived. But the company has been heroic on another level. It became the first enterprise to market a product directly to customers by taking them into a Virtual Reality.

Potential buyers can enter a VPL-designed virtual kitchen in Matsushita's Shinjuku store using VPL's Eyephones and DataGlove. After discussions with the store's expert designers, shoppers virtually tour their creations. If they wish, they can change the design, rearrange the units or bring in different ones. Users can open doors, move crockery, turn on taps and even drop and shatter cups, as they pretend to design their own kitchen. And pretend is the appropriate word. The virtual kitchen can, at best, be described as crude – probably because it is a very early example of a VR world. The graphics remain cartoonlike, the noises bizarre (the removal of a plate is accompanied by a lascivious sucking sound), the hand goes right through doors or walls, and the user floats, making work-surface height checks somewhat problematical. But despite these drawbacks the demonstration is popular and helps the company sell kitchens, if only because of the immense publicity surrounding the project.

Visualising a room in a virtual environment, and being able to change units, sinks and cookers, has implications which go far beyond kitchens. This is the start of something very big. It does not require a great leap of the imagination to see how architects are able to pre-empt difficulties by taking their clients around their houses to see what they feel like in three dimensions, or how city planners can present their ideas to citizens. A late 1992 architectural exhibition in Graz, Austria, will display all such state-of-the-art equipment. In London top-of-the-market estate agents are already using desktop VR to walk clients through their most expensive houses. Interior decorators, builders, landscape gardeners, tailors and dress designers will also be able to help clients by using the technology on their premises. If you are uncertain of your ability to visualise combinations of plants, or new paints and wallpaper, let Virtual Reality visualise it for you – in your own virtual garden or room – and if you do not like it, change it.

Clearly this can be applied to other products. An American upmarket furniture manufacturer uses a VR system to help customers choose chairs, and to introduce an element of personal choice. And why not customise your new car in the showroom? You could add go-faster stripes here, a spoiler there, change the trim or remove the alloys – and transmit the program with your final choice direct to the factory. Why not do it in the carpet store? Just take the dimensions of your rooms along with you. And at your travel agent. Be on the beach in Barbados, walk round a hotel in Honolulu, or a game park in Gambia. Indeed why go away at all? Let it come to you. And as for estate agents – you need never be taken in again by the 'bijou cottage, 2/3 cosy bedrooms, dinky kitchen', routine again.

But why should it stop there? Why not make Virtual Reality home-based? Why not get a virtual catalogue from mail-order companies? Select a dress or shirt, and model it. Size up the lawnmower to see whether it will fit in the garden shed, start it, even mow a virtual lawn. And why stop at mail order? Why not have your local department store update you on its latest products? The store will either send you its latest catalogue tape or disc, or drop its contents into your recorder overnight, through the new cable networks. (As we shall see, other new technologies – recordable laser discs and fibre-optic cabling – will make this procedure much more efficient in the future.) You can then insert the co-ordinates of your living room – the dimensions of all your rooms are held in your computer's memory – select furniture from the catalogue, put it in the living room, move it around, change it, and walk around it to see how it suits. Or do the same with a new bed or bathroom suite.

Again, why stop at these sorts of goods? It's Saturday, shopping day, buses will be overflowing, the car parks will be full and there is bound to be at least half an hour's wait at the check-outs. Why not call a virtual supermarket, browse around its virtual shelves, see what's new, put the goods you want in your virtual basket, feed in your credit or debit card number, sit back and wait for it to be delivered, by old-fashioned butcher's bike! Unlike today's telephone shopping the shopper will be able to see all the goods, and make their choices in the normal way, even down to impulse buying sweets near the check-out. This is not science fiction. It needs a 'home reality engine', helmet, glove, a modem, and a disc recorder (with conversion it could even be done with a conventional video-recorder). W Industries plans to have a low-priced 'home reality engine' version of its arcade game computer on the market within two years, and another leading equipment manufacturer suggests 'Virtushopping' will be in place within the next five years.

There is a downside to this system. All sorts of other organisations could download marketing or promotional material directly into your 'home reality engine' or recorder. Local garages, travel agents, theatres, charities, political parties, time-share sellers entreating you to buy a slice of a virtual villa and even proselytising religions are but a few of the likely sources. Is this a glimpse of heaven? Or is a regular supply of 'Junk VR', all sorts of intense, unsolicited things, descending on you overnight, a vision of hell? Whichever you believe, it will certainly make a giant impact on the way things are done. But to get new VR worlds like these transmitted will need new, cable-based telecommunication links.

When the government decided Britain should be 'cabled', it awarded franchises to small companies on the basis that they would be able to provide cable television, and limited telephone services. This was a very short-sighted approach.

Not unnaturally, the cable companies chose the cheapest option suitable for this type of service – ordinary co-axial cable – which does not have sufficient bandwidth to carry the mass of information that is needed for detailed Virtual Reality transmissions, or indeed other sophisticated interactive business, entertainment or domestic operations. It is possible to 'compress' the information, but even this will not be sufficient over the longer term.

But had BT been allowed to deliver television, we could probably be well on the way to nationwide domestic fibre-optic cabling by now, and VR, alongside other interactive services such as home banking or shopping, would have received a shot in the arm. In addition to fibre-cabling local metropolitan networks BT could have marketed its other services at the same time, and have had an incentive to develop new ones based on information need. However, in the name of competition, BT is banned from television programme delivery. (American telecommunication companies now have a limited role.) A technology push solution, stimulating long-term infrastructure investment, might have been preferable to the demand pull option, especially as that demand has proved to be so elusive. But the cost in other public policy terms might well have been too high.

Some fibre-optic cabling is being done in Britain, but mainly for business traffic. A scheme exists to run fibre-optic cables alongside existing British Rail tracks; US Sprint and British Waterways are putting almost a thousand miles of fibre optics along canals, and BT already has over two thousand miles of fibre cabling in place. But none of these will get Virtual Reality – or interactive services – into small businesses or homes. That last stretch, into side-streets and ordinary houses, flats, offices and small factories is too expensive given the legislatively restricted returns. The Americans are little better, but with some experimental consumer areas. Cerritos in California has a broadband network linking two schools, and residents can call up videos on demand. The Mississippi 2000 project provides full interactive video to four schools, hundreds of miles apart; and both NYNEX in

Boston and VISTANet at UNC provide broadband fibre-optic links into hospitals. There is the cabling of Bellingham by US West, and one Californian district, La Crescenta, is testing squirrel-proof cable. But the majority of American fibre-optic cabling is confined to academic, government and military uses.

Very few technologies of any worth remain underused for long, and Europe clearly believes fibre optics has considerable worth. The entire European cable infrastructure is due to be replaced by fibre over the next twenty years at the cost of several hundred billion ECUs. As a start, HERMES, an unlikely consortium including European railways, Daimler Benz, Compagnie de Suez, NYNEX and US Sprint, plans to build a high-speed fibre network linking twenty-four major European cities. But France is leading the way. Biarritz is fibre-cabled and the aim is to have fibre cabling in seven million homes by the end of 1992, although there appears to be some slippage. Even so, today there are French dwellings with enough cable bandwidth to allow Virtual Reality programs to be piped straight into their living rooms. 'Home reality engines' will not be needed. VR services will be delivered at any time of the day or night directly to the user, or into storage on a recorder. Furthermore, as it is an interactive system, the user will be able to contact the sender immediately – and in Virtual Reality – if that is appropriate.

But if the French are ahead in Europe, in Japan telecommunications has the status of a national priority. In March 1990 MPT, the telecommunications ministry, and NTT, the telecommunications giant, announced their Visual, Intelligent and Personal Services plan. It will fibre-optic link *all* Japanese homes and businesses by 2015 at a cost of one hundred and fifty billion pounds. If this sounds a lot of money, so is the return. By 2020 the Japanese estimate that thirty per cent of their Gross Domestic Product (GDP) will come from broadband links. Much of this will come from videotelephones. NTT estimates there will be over five million videophones in use over the next twelve years, but like

the initial calculations on the penetration of colour television this might well be a substantial underestimate. (By
1997 ninety per cent of central Tokyo business buildings
will be connected by fibre.)

Given their lead it is probable that French, Dutch and
Japanese companies will capture most of the market for
domestic and business appliances that will use fibre optics.
Yet again Britain and America will have to import a new
technology to satisfy their own consumers, a technology that
they themselves did so much to develop.

There is an alternative method of transmitting Virtual
Reality signals, which BT is intending to use. It is possible
to use radio links. However, these have to be very short
wave and high frequency to get sufficient power, and this
may give fading, interference and shadowing. It might, however, prove to be the best method of getting Virtual Reality
from the kerbside into homes and smaller businesses. Fibre-
optic cabling will be used for the trunk routes, while radio,
placed, say, atop of lampposts, could beam the last stage
directly to consumers and other users.

Virtual Reality itself does not need to wait for fibre-optic
cabling. The VPL system which created the virtual kitchen
has been used to produce a far more realistic graphic world
for a Berlin-based multimedia centre, Art+Com. Since the
destruction of the Wall the Potzdammerplatz underground
station, bricked up and unused since the end of the war,
has become a symbol of Berlin's reunification. There are
plans to reopen it, and because of its importance great care
has been taken over the designs for its modernisation. The
VR system was used to help evaluate the different design
proposals, in the round, as it were. Because the graphics
were crisp, sounds realistic and movements controlled, it
was possible not only to test practicalities (whether platforms were wide enough for example), but also allow the
public to sense what the station would feel like. Ambitious
plans are now in train to model the entire city of Berlin
using Virtual Reality. A desktop VR system has been used for
a similar walkthrough, modelled by Cadommac, for London

Underground and Cross Rail at King's Cross in 1991. The company call their method Virtistic, and have used it to model an art gallery and underground pipe networks, among other things.

The first recorded use of Virtual Reality in design was, appropriately enough, at the University of North Carolina's new computer science block, Sitterson Hall. After more than a little scepticism and hesitation, the architects agreed to 'tour' a virtual representation of their proposed building, using a helmet and a steerable 'treadmill' for steady, non-flying progress. Once they were able to walk around their creation, they had to agree with the faculty that one of their partitions was in the wrong place, obstructing the normal flow of people. The design fault had been missed using traditional methods, and the architects were able to amend the building *before* it was built.

The detection of design errors is, of course, a major plus for Virtual Reality. It is far easier to recognise potential difficulties, or actual mistakes, when moving around inside a design, rather than looking at two-dimensional plans from the outside. Most instances of Murphy's Law (if it can go wrong, it will) are caused by a lack of sufficient preparation or forethought. VR provides the opportunity to do the one, and exercise the other. Furthermore, it is possible to repeat the virtual tours until there is a satisfactory outcome. Mistakes can be rectified before they are actually made, or indeed become fatal.

Design goes far beyond architecture; indeed it affects most of the physical artefacts that surround us. Costly errors often occur in the course of the design process. John Thomas, Director of the NYNEX Artificial Intelligence Lab sees Virtual Reality as helping to save some of the astronomical sums of money lost through bad design. He also argues that it will allow designers and architects really to experience other people's problems. Slim architects will be able to appreciate the difficulties of fat clients in narrow doors, or of wheelchair-bound people with conventional kitchens and toilets. Understanding other people's needs in this way is just one part of our general need for good communication networks. Communications make the world go round. People are gregarious beings, they need to meet, talk, be praised, chided or stimulated. Both hermits who escape from the world and Trappist monks who believe that their vow of silence is a sacrifice, make the point eloquently.

From the moment he started with Virtual Reality, Jaron Lanier saw VR as an evolutionary development of the telephone, rather than a new type of television. The distinction is clear. You take part in telephone conversations, but you receive television. This view led him to pursue the idea of the virtuphone (Virtual Reality videotelephone) where you contact people using the telecommunication system, and share the same virtually real space. This attractive proposition will give free rein to the imagination. You may choose to represent yourself as a cat or whatever, and as Lanier puts it, 'You don't have to say you are in a bad mood, you just produce a thunderstorm.' The concept has been proved in research studios in California, has been demonstrated to Bell Corp executives, and it works. But for mass public use it awaits widespread fibre optics.

However, the do-it-yourself side of Virtual Reality may need nothing more than an ordinary personal computer (PC). Both Sense8 in America and Dimension International in Britain sell their virtual world kits to research facilities, universities, companies, and police forces. Neither Sense8's WorldToolKit nor Dimension's Superscape kits are cheap, costing well into five figures, but both are increasingly being seen as excellent visualisation tools. Indeed, the leading British Virtual Reality consultancy, Virtual Presence, claims there are two hundred establishments in Britain using Virtual Reality DIY kits, among them departments in almost all the universities. Bob Stone of ARRC agrees with the estimate but suggests that too many of the university researchers have no real idea what they are using VR for.

Interestingly, neither product needs conventional programming skills, although WorldToolKit requires an ability to handle AUTOCAD, and Superscape needs a person who can handle menus and icons. Dimension International also produces what must be the world's cheapest VR product. For less than the average price of two West End theatre tickets, their 3-D construction kit (in America known as VR Studio) sells to computer games buffs. Although it is intended for games players, it is a genuine real-time three-dimensional product, and as Dimension's Ian Andrews says, 'It's a wonderful way to start learning about VR.' As of now most of these systems are running on a two-dimensional screen, but all are capable of being transferred into fully immersive VR. Once the visual drawbacks of the helmet have been overcome and prices fall, as they will, immersion will be the order of the day. But until that time, probably four years away, big screen, desktop and Kreuger-style VR systems will be the more practical.

Jonathan Waldern of W Industries initially saw demand for Virtual Reality in terms of engineering, medicine, and helping the handicapped. Instead, arcade games have proved to be his major marketing opportunity. But the Expiality computer that drives his games also drives the work station test-bed that W Industries sells around the world, and as we have seen this will shortly be available for home-based VR games. But just like ordinary PCs it can also be used for other programs, and with the Dimension International and Sense8 'toolkits' providing the bricks, mortar and polygons, the stage is set for the first immersive, self-designed, self-created virtual worlds.

Who will use them, and what will they try to make? No one knows. Perhaps people will start to take design seriously, and attempt to change their own houses and gardens; or try interactive art, or movie making – perhaps with animation. They may try to model the processes of their own small business or club. Will they devise relaxing environments to use at the end of a hard day's work? Or will they, as is more likely, design stimulating worlds: do-it-yourself karaoke backgrounds, perhaps? They might just invent a fantasy world. And will their worlds be shared or passive; constructive or defensive? Will the old acid-heads find a new place in which to hide, or will DIY virtuality replace ecstasy for the new generation? It must be assumed that people will design their own heavens, and unlikely that many will design deliberate hells; although one person's heaven may well be another's hell.

Despite the number of home reality kits sold, the idea of making your own world has not reached a wide public consciousness as yet. So how will Virtual Reality become a mass consumer good? After all it will not be cheap initially. It is highly likely that the engine of awareness, and fastest growing sales, will be the entertainment industry, both for in-home and outside use. Price will be as important for mass sales as for arcade owners, but not necessarily at the very outset. Fortunately, a few consumers buy products just because they are new; economic jargon calls them 'positional goods'. You probably know one or two people like them. They wore their Walkman before you knew it existed; and a CD player before the media hype, and now have programmable videodisc players. These are the people who will buy the first 'home reality engines', even if (or because) the price is high. They set the fashion and pump-prime the market. As we explained earlier when new products are sold and positive cash flows and profits are generated, prices fall and quality rises. In this respect Virtual Reality would appear destined to follow in the footsteps of ball-point pens, colour TV sets, electronic calculators and, more to the point, personal computers.

There is, however, a strand of Virtual Reality which is both refreshingly radical and altruistic, indeed almost anarchistic. It maintains that information should be shared freely. A company rejoicing under the name of The Stone Soup Company has provided the source code of their program called Fractint free to everyone. It enables users to produce their own fractals with a vast choice of colours, in a state-of-the-art way, and runs on a very wide choice of computers and monitors. In January 1992 Dave Stamp and Bernie Roehl started to give away the source code for their program for creating a three dimensional computer environment on

Rend 386 and 486 machines. All you have to do is log on to the CompuServe network and find the library in Room 13 of the Computer Art section, and hey presto you have the tools necessary to fashion your own virtual world; free, gratis and for nothing.

But these are the exceptions. And while price will become very important, it is foolish to try to estimate guideline prices today, either for consumer or other Virtual Reality systems. To say a home reality engine should cost under fifty pounds or two hundred dollars is to court the furies. While, obviously, the lower the price the greater will be the number of systems sold, the 'correct' price will depend on many factors, including a country's inflation rate, its disposable income distribution and other economic parameters, the competition for the product, the state of consumer confidence and, above all, what is fashionable at any one time. Today, fashion is as important in electronics as it is in clothes and cars. It is often set by the demands of young people, and most often created and stimulated by television. If the fashion at Christmas 1995 is for home Virtual Reality games, then nothing will stop it.

Virtual Chapter 5

The world is full of Doubting Thomases, critics who damn every new idea with either unrelieved gloom, or the faintest of praise. Those who believed VR was merely hype must have been astounded by the City's reaction to Division. This small company, which moved from the charm of Chipping Sodbury to the bustle of Bristol, and now sports an American off-shoot, became the world's first dedicated VR company to get a full Stock Exchange listing. On 17 May 1993 it opened in London at 40, went to 84 in three minutes and closed at 103. Even hard-headed financial institutions believe in the future of VR.

And with good reason. John Hough, managing director of the London-based VR consultancy and equipment supplier Virtual Presence, believes today's VR market is around $50–

75 million, excluding the entertainment and military markets, which he admits would more than double the total. But this is a technology of the future. One large US telecom company estimates that by 1999 the market for VR-based information design will be more than $1 trillion a year. As Bob Jacobson of WorldDesign points out, 'Even if this is wrong by a factor of a hundred, the potential market for this technology is impressive.' 4th Wave, a US multimedia consultancy is slightly less bullish. It estimates the late 1990s market for VR technology to be $1 billion in business applications; $600 million in consumer applications and $500 million in R & D.

Although the future may be bright it was a mixed year. Those companies which stayed in the race, such as Virtual Presence, have done well. Its turnover increased by five hundred per cent, and it now has over fifty clients, mainly British, although both German and Dutch companies figure strongly. Bob Stone at ARRC created a consortium – Virtual Reality and Simulation (VRS) – with nineteen companies including ICI, Rolls-Royce, GEC, and BNFL collaborating. As part of the arrangement, the University of Salford's Surveying Department created a Virtual Reality chair for Stone, who becomes the second VR professor (just behind Roy Kawalsky of BAe at Hull University). It is not purely honorific. Stone and colleague Andy Connell will teach VR to third-year degree students at University College Salford.

But not everyone did well. Jaron Lanier's VPL, the world's most famous VR company, filed for bankruptcy after a Singaporean funding disappointment and consequent financial collision with major shareholder Thomson-CSF. Although the company still exists it is unlikely to be viable. Users with VPL equipment were unsupported, and bought from rivals who moved into the good market opportunity. Lanier and other VPL refugees have opened a new VR software company, Domain Simulations, while the rump of VPL is engaged in patent law-suits. This is the first – but is highly unlikely to be the last – high profile disaster.

The VR conference market, however, goes from strength

to strength. There must have been at least forty over the past year, including specialist conferences on medicine and the disabled. The European Commission's Directorate XIII C convened a conference in spring 1993 which gave birth to a small expert group looking at specific VR applications in education, training, health, the disabled and transport. This has now moved into Phase II with detailed work plans to isolate the most useful areas of application, the research to be conducted by multi-disciplinary teams. Outside Japan it is by far the most ambitious government approach, dovetailing as it does with the EC's own ESPRIT fundamental VR research programme. One of the authors is involved in the EC study and formulated the following rules governing the use of VR. *All* four must be met to justify VR use.

Rule 1 VR should be used only when the application *cannot* be done in a natural environment, *or* would be so expensive/hazardous etc that the need should not be met in a natural environment.

Rule 2 The cost of the application must be within acceptable market or cost/benefit bounds.

Rule 3 The technology itself must be reliable, robust, capable of being serviced and meet rigorous health and safety requirements.

Rule 4 The use must be widely socially acceptable, and conform to the ethics and norms of any involved professional bodies.

As we suggested a year ago it is still too soon to expect a flood of finished VR consumer goods or services. But marketing and advertising have continued, often using W Industries machines, attracting media attention in the US, Japan and the UK. However, a VR system taken to the 1992 Conservative Party Conference proved a complete failure as only one politician was brave enough to try it!

Current research in VR design, manufacture and visualisation may have beneficial spin-offs for consumers in the not too distant future. Projects sponsored by Rover/BAe on VR-assisted manufacturing, by Chrysler on design and by Renault/Volvo on visualisation will speed up personal car order delivery times. Daimler Benz is evaluating VR as a method of simulating 'active safety' in its cars, especially in the psychological aspect of their interiors, seats and instrument layouts. A large supermarket chain is actively working on VR as an aid to warehousing and distribution, and in Holland the Calibre Institute of the University of Eindhoven has created a complete virtual town using WorldToolKit for architectural purposes. Indeed, more architects are using client walk-throughs. The company providing these, VRTD, also uses a desktop VR model of the proposed redevelopment of Newcastle-on-Tyne's waterfront for consultation with local user groups, consumers and possible clients. On the same track the University of North Carolina (UNC) is engaged in modelling the Los Angeles riot areas to aid reconstruction, and German architects are constructing a virtual new East Berlin, helping to rebuild the real one.

Home VR has taken several steps further forward. SEGA's announcement of a low-cost (about two hundred dollars for the headset) VR game system has taken the industry aback. If VR games can be put into the home at this low cost, clearly all sorts of other VR for marketing, politics, religion, pornography or education and training will not be far behind. A second step has been taken by Phillipe Queau of the MediaLab in Paris. He demonstrated a televirtuality link between Monte Carlo and Paris on ordinary telephone lines. A priest in Monte Carlo wearing an HMD watched his own clone (or puppet) taking a woman in Paris (who saw the same thing) on a guided tour through a splendid virtual vaulted halls of the Abbey of Cluny. It did not need fibre-optic cables as only the changes to the virtual world were transmitted. This technique opens up a whole new vista of VR use at lower than expected cost. Finally there is a great deal of work being done on both videophones by BT and fibre-optic networking and compression, especially at the Swedish Institute of Computer Science and by ARPA in the US. The infrastructure necessary for home VR is getting nearer to being in place far earlier than we anticipated twelve months ago.

6

Industry and Commerce

It is one thing for home-based consumers to be seduced by fashion, but quite another for the corporate sector. There was a time, in the 1960s, when companies rushed out to purchase mainframe computers just because a rival down the road had one. The resulting 1970s débâcle stopped this practice dead in its tracks. And although there appears to be an almost universal fashion for companies to move into the most overpriced and over-officed headquarters, the basic lesson appears to have been learned. Companies tend to stick to the rule that they should use a technology only when it is highly likely to create profits and sales, improve quality and customer care, and cut costs.

The 1980s was a decade of intense industrial change. Automation in the form of 'smart' machines, robotics, computer control and just-in-time systems were introduced into almost all manufacturing processes. At the same time, service industries became heavy users of computer power, especially in areas impinging on the public. Entire sections of society became dependent on computers; government as much as supermarkets and banks in the same way as oil companies. But in this period computers themselves went through different stages of evolution, culminating in personal computing, laptops, and executive information systems. Each new product took some time to be appreciated, and then used widely. Virtual Reality will be no exception to this rule.

The aerospace industry has certainly recognised the value of Virtual Reality, albeit predominantly on the defence side. For example the LHX simulator developed at NASA/Ames is claimed to have saved the American army around $1 billion on the development of its light helicopter because different prototypes did not have to be physically constructed. However, there are signs that VR is now being used in civil aircraft production and design. Boeing maintains its own VR design facility, as well as operating on the HITLab's VEOS system. The company uses them to help design cockpits, flight cabins, seating patterns, safety features, cargo spaces, service points, and other passenger and crew requirements. As far as Boeing is concerned it is a proven technology; it commissioned Meredith Bricken to design a virtual rotor-plane, the VS-X, which makes virtual test flights – over surreal countryside. From Boeing's viewpoint it is both a money-saver and a customer-preserver. It helps iron out irritating little design faults, like meters obstructing inspection hatch views. Mechanics, their employers and airlines are the better pleased because service jobs get finished more quickly, and with less risk of error. Boeing is pleased because it gets the credit, and sells more planes. Hughes Aerospace and BAe also use VR systems, Hughes in the maintenance areas (and through its ownership of Rediffusion) and BAe in its multi-site Supercockpit work.

However, in general, manufacturing industry is only just catching up with VR. SATRA, the footwear industry's research and consultancy centre, is using a WorldToolKit as an engineering tool to improve manufacturing methods. And although Caterpillar has used it to amend its tractor driving positions, most interested companies, like the BT and GECs of this world, are buying VR test-beds and kits to see what they can be used for, rather than with fixed production or design plans in mind. While GEC is almost certainly interested in the defence sector – it will not divulge anything about its extensive VR rigs – many other companies could, on the face of it, use the technology to considerable effect.

The Germans have realised this and have opened the

Centre for the Demonstration and Competence of Virtual Reality in Stuttgart. The centre, which is open to businesses both large and small, and acts as a consultancy as well as a showplace, has three sections. The graphical data management section is of as much interest to government, universities and the service sector as it is to manufacturing industry, while both service and manufacturing companies find the ergonomics and control section, run by IAO, very useful. The Fraunhofer Institute for Production Technology and Automation (IPA) runs the design and production section, which clearly is of most interest to manufacturing industry. The IPA reports an increasing level of interest among German manufacturers – including smaller companies – especially in robotics and telepresence. Both the institute and its client companies are also hopeful that the three-dimensional, interactive VR design tools will bring the long-awaited breakthrough in simultaneous engineering, where design and production run concurrently. From the authors' point of view the long-awaited design breakthrough would be as simple as a car with an accessible dip-stick.

Design was once a very under-regarded area; think back to the sharp edges under car wings, television pictures that could only be adjusted by double-jointed acrobats, and cookers with uncleanable ovens. But now, because consumers are more demanding, safety standards higher, and competition fiercer, design is coming into its own. The 1980s growth of consumer organisations signalled the enfranchisement of the grumbler, and manufacturers have had to respond.

Whether designers or engineers are working on space-stations or toasters, Virtual Reality gives them an extra dimension – in all respects. An engineer designing a new vacuum cleaner in real-time VR can walk around the virtual machine, change parts by handling them, and generally do what an engineer needs to do, not what a computer program allows, which all too often has been the case. Furthermore, the engineer can 'be' the consumer, pushing different designs around, trying to change the bag or put it into a cupboard, without the trouble and expense of building several prototypes. It is a financial, creative, commercial and professional release.

But why should the engineer alone attempt to change the virtual cleaner bag? Surely it would be better to invite consumers to try for themselves? And if public reaction and opinion is important, market research in general could be revolutionised by testing proposed products on the public in virtual form, or by using virtual supermarkets rather than asking people to imagine them. Consumers would then become part of the design process in a far more realistic way than at present. Another VR manufacturing use being researched at Boeing in Seattle is an adaptation of a pilot's 'head-up display'. Operators wear goggles which superimpose a three-dimensional image of the ideal solution of the job in hand, perhaps installing piping, wiring or machining a widget. It increases both accuracy and speed, as operators no longer need to consult manuals all the time.

The field is limitless; from television sets to space ships, and from trucks to cornflake boxes, VR will be standard design technology. The IBM announcement of its first CAD Virtual Reality work station in March 1992 shows that the technology is starting to come of age. An IBM RS 600 work station will be combined with Division's VR hardware and software. The system will take existing CAD systems and convert them into three-dimensional, interactive, helmet-based virtual worlds. The first Silicon Graphics Computer Aided Design (CAD) packages, which enabled engineers and draughtsmen to operate directly with computers were a breakthrough; Virtual Reality is a revolution of equal magnitude.

And as the technology becomes cheaper so the number of users proliferate. In January 1992 VPL launched their MicroCosm VR system running on a MAC and complete with DataGlove and Eyephone, all intended for desktop use at a very affordable price. This type of machine opens the door to VR use in all sorts of design, for example new traffic control and road systems, theatre and film sets, or new tax regimes. Virtual Reality puts you, the designer, in the position of driver or pedestrian, tax collector or payer, whichever perspective you need to take. You can find out what it would be like, if and when you decide to make the changes, and you could, if you were confident enough, invite the public to participate as well. Research and prototype model building for uses such as these are under way in Berlin, Nottingham, Japan, and of course in California.

The pharmaceutical manipulations being done at the University of North Carolina, with Professor Fred Brooks in charge, represent the other end of the VR spectrum. In many respects Professor Brooks, whose interest in Virtual Reality stretches back to the early 1970s, is the doyen of VR academic research. His department has been funded by various American government bodies, as well as the drug industry in the shape of Burroughs Wellcome. They are well past the research stage in several areas, the best documented being the manipulation of virtual molecules of pharmaceutical products. This system uses a force feedback arm which allows researchers to 'feel' whether the molecule and its under layer are repelling or attracting each other, and with what strength. These electrostatic forces are complemented by histograms of the energy stability on the Pixel Planes screen. The result is a visual experience complemented by a physical sense of 'feel' (known as 'haptic perception' in VRspeak) of a real solution to what is generally an abstract problem. As it speeds up the development of new medicines, it is commercially viable.

Commercial considerations lie behind similar work being done at the University of York. Glaxo, the British pharmaceutical company, is funding a project using IBM and Division equipment to visualise, model and manipulate chemical and protein structures. Quite how this differs from the UNC systems is not yet apparent as the project is currently under wraps. But although it is early days Glaxo is confident the technique will enable it both to speed the discovery of new medicines and help solve technical problems, such as difficulties with crystallisation.

It may be pharmaceuticals today, it could be almost any other molecular combinations tomorrow. Is this the techno-logical breakthrough that both genetic engineering (at a molecular level), and nano-engineering (at an atomic level) have been waiting for? Is it the 'lathe' on which new plastics and biochemical industries will be started? Certainly the ability to visualise and manipulate the unseeable can only be of assistance for future developments of this type. There is a growing number of scientific areas where VR is now starting to make a real contribution.

Viewed from the outside science appears to be dominated by mega-expensive equipment, strings of equations and chains of data, all linked by rigorous thought patterns. So it is a touch surprising to find eminent practitioners claiming that the use of sight, touch and intuition gives them a better understanding of technical and conceptual problems than conventional data. But perhaps, on reflection, it should not be so startling; half the human brain is devoted to processing visual information, and VR is the first computer system to make full use of it. Scientists can be in the computer loop themselves and see, or in the case of UNC, also feel what is happening. The fact that scientists need no longer try to visualise what their data means is a development of some magnitude. It is as though they have been given another sense, or a third hand, to enable them to grapple with increasingly sophisticated concepts.

Most scientific research is funded in the hope that ulti-mately it will lead to a commercially viable product. Any-thing that helps in this process is important, and that is why visualisation is seen as a giant step forward. If you are deal-ing with the invisible – magnetic forces, electron densities or air turbulence – an aide to 'seeing' what happens is more than useful. And the three-dimensional modelling of forces in physics: fluid dynamics, crystal formation, thermal maps, stress patterns and fluid flows, can be vital in understanding the nature of these subjects. Now, for the first time, these forces can be made transparent in VR, allowing all the scien-tists' faculties full rein. Visualisation is also being extended to less arcane subjects such as astronomy, geophysics, seismology, and meteorology, where even now we can visualise, and reproduce in a television weather forecast, the movements of low-pressure fronts, clouds or hurricanes. While what we see on TV is not Virtual Reality, in American meteorological laboratories it is possible to 'ride' a storm, or slide along an isobar.

Virtual Reality has the immense, and so far untapped, virtue of making difficult concepts or constructs easily understood, so that the relatively unskilled can make sense of them. For example, weather forecasting can be so well presented that explanations from professional forecasters will be irrelevant – except of course for the blind. And there are many other opportunities. Suppose a complex telecommunication system in a large office block goes down. Nowadays, small diagnostic programs called 'knowbots' can be placed into the system to detect faults and report back. If a VR model of the system was constructed floor by floor, a user could navigate the entire system by 'becoming' the knowbot. Systems could be monitored, faults pre-empted or traced more quickly, cheaply and accurately, using the services of relatively unskilled personnel.

Air traffic control is another prime area where VR can be used effectively. Using Virtual Reality to convert their two-dimensional screen into a three-dimensional space, controllers can 'be' in space with their charges. By touching a virtual plane with their glove they can put it on the appropriate flight path. A screen in the pilot's cabin shows what has happened, backing up verbal commands. While initially the VR system will be used as a teaching tool, future controllers will no longer need to interpret radar and voice communications. And because this system is based on everyday actions the chance of errors should be smaller. But because 'reading' radar will not be crucial, an element of job de-skilling is involved, even though the controller's weighty responsibility remains the same. Science fiction? Not at all – they are working on VR air traffic control at Brooks Air Force Base in Austin, Texas.

Many people are intimidated by collections of abstract statistics – incomprehensible, dry-as-dust rows and columns of figures and graphs – and lack the fortitude and patience needed to explore them. These negative feelings can be overcome by putting the data into a visual, well-understood analogue form. Californian academics are now exploring the visualisation of large abstract databases typified by crime, court, business and economic statistics. Such an approach is rare. Professor Philips once made a working, hydraulic macroeconomic model, now languishing unloved and unused at the London School of Economics, but that is about all.

As things stand today, we have to rely on statisticians (or worse, politicians), accountants and economists to provide and interpret government data, balance sheets and accounts. Without being experts ourselves this is the only way we can understand such technical matters. But they may also affect our everyday lives, so that democracy itself may depend crucially on the understanding of this information. Regrettably, all too much of it is presented without thinking of the consumer, and as a result few people understand, or even attempt to understand, the messages. It is like presenting a scientific paper on thermodynamics to people at large, rather than putting it into the BBC's *Tomorrow's World* format. But Virtual Reality can solve the problem. It puts data into easily understood analogue forms. As cognitive research is applied to VR presentations, more people will be willing to approach the information in the first instance, and then understand it. In this, and many other direct ways, VR has the potential to change politics like no other technology since television.

To be more precise, computer graphics and multimedia have prepared the way for this; Virtual Reality makes it more vivid, by putting users inside the information. Companies are starting to make use of this property. Several companies have contacted the Centre for the Competence and Demonstration of Virtual Reality in Stuttgart to find out how graphical data-management could help their business, and some have been impressed sufficiently to embark upon feasibility studies. Approaching the problem from another angle, Dimension International has sold its Superscape to a very wide range of users. One needed to visualise storage arrangements in warehouses, BT needed to visualise electronic data, and civil engineers and scientists in both Britain and Italy are customers. Ian Andrews, Dimension's managing director, is also proud of the fact that the company is selling the product to large Japanese corporations. He believes that Japanese executives are far further down the road of thinking in visualisation terms than their Western counterparts – which may be worrying.

Specific tailor-made visualisation systems are also being developed. American software houses like Precision Visuals of Boulder, Colorado, have been putting stock market and financial analysis data into three-dimensional visual forms. But as a Wall Street analyst claimed recently, 'A static, out-of-date visual is as useful as a freezer in an igloo.' He pointed out that whatever graphics are produced nowadays, they must be updated in real time. And it is only a short step to turn these CAD graphics into VR, so that analysts can alter one or more parameters, and then sample and manipulate the mechanisms of the consequential changes from the inside. And only a short step from there to a virtual space in which analysts, dealers and brokers can argue together inside the information!

But Dr Charles Grantham of San Francisco University has taken the idea a step further. He has produced a VR model of a fictional business in full flow, using valves, dampers and cylinders as metaphors. Although not immersive, there is no reason why it should not be, especially if managers want to know what happens when price, output, quality or whatever, alters. Although these answers are straightforward to estimate conventionally, the end result tells you little about the dynamics (the process), and at what point intervention would be counterproductive, or useful. Business VR visualisations may be the ultimate management consultancy tool, especially for those peddling nostrums based on 'right first time'. (VR allows you to make the mistakes first – painlessly.) Alternatively, it may put most of them out of business. Even the weakest managers will be able to grasp what is happening with VR models on Grantham's lines, and as they can be run on desktop two-dimensional screens, costs can be kept low.

While it is also possible, if laborious in the extreme, to model towns, regions, countries, continents, oceans, even the entire world, the fascination of VR lies in its ability to model believable representations of the unknown. NASA has used this property. Electronic signals sent back from the *Viking* Mars orbiter have been converted into computer graphics, building up three-dimensional pictures. With the appropriate VR equipment it is possible to 'walk' on its surface, even with the correct gravitational pull. And if there is dense cloud cover and an ultra hot surface, as on Venus, a radar-equipped spacecraft, like *Magellan*, can send back 'pictures' with the same results. Overall the space industry finds Virtual Reality a more than useful tool.

Repairs (or engineering work) in space can be a problem. Astronauts may be unavailable, the repair may be outside their sphere of competence, or it may be too dangerous or inaccessible. This can be overcome by using remote telepresence robots, controlled by operators equipped with head-mounted displays and gloves. One method is to use 'virtual icons', but 'virtual puppetry' is more spectacular. The operator raises a hand, the robot raises the equivalent piece of its machinery; if the operator's hand tightens to a grip, the robotic clasp will grip whatever is in its path. With force feedback the operator also will 'feel' what the robot is holding. If the operator looks left, the robotic visual sensors move to the left as well. But the vital point is that the operator does not see the robot, only what the robot 'sees' through its three-dimensional camera 'eyes'. This can result in the distinctly off-putting 'out-of-body' experience, where you look straight ahead, but see your virtual self in profile!

If visibility is poor, perhaps from fire or chemical smoke, the camera image is transformed into graphics on the operator's head-mounted visual display. So for all practical purposes the operator is where the robot is. While astronauts in a shuttle or space-station operate the robot, it can also be operated from other places in space, or from Earth.

Although components need refining, the technique is proven, and has far-reaching possibilities. The Japanese are putting much effort into telepresence (it is rumoured they wish to develop their own space capability, with a view to building space hotels). The IPA in Stuttgart specialises in telepresence robotics, using Puma models, and have now reached the point of marketing the world's first off-the-shelf telepresence robots. But the first successful graphic telepresence demonstrations were made at the Advanced Robotic Research Centre in Salford, where Bob Stone and his researchers, working on a tiny budget, are looking at a host of possible applications. (These include research and development in several areas with the European Space Technology Centre in Noordwijk, Holland.) Clearly, the computer graphic option is favoured for restricted visibility work. Robots can be equipped with radar, infra-red sensors, geiger counters or X-rays. Uses will almost certainly include undersea work in turbid water, in firestorms, smoke, chemical spillage and fumes, and total darkness. Fire-fighting, bomb-defusing, space-engineering, security, 'dark' factory or computer-centre maintenance, even remote computer loading or armament stacking robots are perfectly possible, indeed likely in the years to come. But even today VPL has just proved the technology by producing a telepresence robot for the Japanese construction company Fujida, which was operated on a cross-Pacific Ocean link. The distance, and the fact it used a non-fibre-optic cable link, gave it a two-second lag, but its graphics and its functions all worked satisfactorily in murky, hostile conditions.

Telepresence robots get to the places real people cannot get to – such as highly radioactive areas. One of the most urgent tasks in today's world is to repair and decommission nuclear installations. Eastern Europe, from Russia to Bulgaria, is littered with monuments to Soviet technology, decaying nuclear power stations crying out for emergency servicing. But although there is little or no money available to do it, we ignore the lesson of Chernobyl and bad design at our peril. That disaster also showed the fallibility of conventional robots in radioactive and smoke-filled environments. But, by definition, telepresence robots cope with a lack of visibility. They can also be armoured against radiation and made as cheap disposables. Either or both options will give short, intense periods of work. Some robots may be mobile, with their own anti-collision sensors, others fixed. A judicious mix could get robots into 'hot cells' and remote operators could close down or isolate them. Although sited in Eastern Europe, the repercussions of an incident would be pan-European. And we should remember that many reactors nearer home, in Britain, France and Austria, are ageing and will need to be decommissioned in the relatively near future. Both the ARRC and the IPA in Stuttgart (among others) are ready and awaiting the call to put their robotics to work. ARRC is also involved in helping NYNEX to dispose of spent nuclear fuels, using both robotics and simulations.

Simulators have been with us since the Link trainer, well before nuclear power stations. Virtual Reality is the most recent of the line. Conventional physical simulators can be expensive, be they for pilots, divers or rides at Disneyland. Simulating weightlessness in the space industry can cost an aeroplane, flying time and fuel, and then you get only seconds of weightlessness per barrel-roll. It is far easier, and considerably cheaper, to take the astronaut into a virtual weightless world; moreover, one that is natural to Virtual Reality (it saves programming the gravity). It is also cheaper to model a virtual capsule, buggy, shuttle deck, or control room, than build one for training in ordinary and emergency procedures. NASA is some way down the road to this destination, some way ahead of the Europeans. Interestingly Bob Stone of ARRC has been contacted by the old Soviet space people in Star City with a view to helping them evaluate Virtual Reality.

In fact, Virtual Reality should be able to produce unparalleled simulation experiences. Leaving aside the entertainment, medical, military and education sectors, which we consider in the following chapters, VR can be used to simulate conditions for almost all forms of activity. If workers at the University of Valenciennes are working on making virtual explorations of long-lost monuments and buildings, why bother to take visitors on tours round today's factories when you could do that virtually as well? And think what it might mean to driving schools, diving supervisors, travel and tourism, train-drivers, control room operators, visits to libraries and museums, hobbies like antique collecting or tatting, and all forms of sporting activities, from skiing to bowls, and from baseball to soccer.

Japanese telecommunication companies have not only seized the technology, they have grasped part of Lanier's telephone philosophy: the idea of teleconferencing. As we have seen NTT is putting great store by the videotelephone, and teleconferencing extends this from two people to a 'meeting' between any number of people at different locations. Each participant will see the others on small television screens. Introduced in the 1980s it was not enough like the real thing to be a success. Discussions were stilted, and participants' reactions difficult to evaluate. Furthermore it was expensive. In the mid 1980s it cost a thousand dollars per hour, calls had to be reserved days in advance, and it needed half a million dollars worth of equipment at both ends. Not exactly a commercial proposition. You could ferry four executives from Los Angeles to New York almost two hundred times, and still save the running costs! But today, two digital lines will carry a spontaneous conference for fifteen dollars an hour, with (shortly) only ten thousand dollars worth of terminals at either end. It is starting to make economic sense.

But this still leaves the unsatisfactory nature of the conference itself. Several Japanese companies are working on transforming video into very high-grade graphics, concentrating on the most difficult bit, the face. They could then run a teleconference in a shared space: in other words everyone would appear to be sitting around the same table. Using spectacle screens, person A would only see persons B, C and D; person B would only see A, C and D and so on. Personal cues would then dictate who was to speak next, and body language reinforce status. The Myron Kreuger Videodesk version of VR is very similar, and although his system concentrates on silhouettes, hands and desktops, it has interested British Telecom sufficiently for it to send a team to check on it. Mandala, based in Toronto, also places people in a shared space, but in this case its video images are seen inside a shared virtual room. This may well be the model for future teleconferencing, and almost certainly telecommuting.

Bob Jacobson, who worked for some time in the Seattle HITLab, has postulated the concept of the Information Environment, which he believes will be composed of synthesised sights, sounds and tactile fields. But he does not underestimate the difficulties involved in designing these environments, in which he believes future information systems will circulate, nor indeed the problems associated with the virtual teleconferences which will take place in them. On the one hand there is no natural order in which the environment should be designed, and on the other there is the difficulty of designing the software and sophisticated switching systems which will enable people in a teleconference to get to speak in their proper order. While he believes neither difficulty is insuperable, he knows that much research, and then development remains to be done to bring this part of VR near to its potential.

Japanese telecommunication and computer companies are roving over a wide range of fields, stimulated, co-ordinated and part funded to the tune of almost seventy million dollars by MITI. Sony is developing wearable computers, and new forms of distributed operating systems. NTT is trying interactive systems which will react to the user's expressions and body language, as well as voice. Universities, institutes and smaller research companies are engaged in similar fields as well as in telepresence and robotics. To the outsider it appears that a national priority goal has been set, and subdivided into smaller tasks – rather like parallel process transputers. Some researchers work at colour transmission, others at graphics, at tracking, at three-dimensional video, at boundary detection, at tactile simulation, and at standards, while yet others are attacking the spatial computer interface that can be navigated by the user. Because of this planning, the thorny problems afflicting America in particular, of research overlap and duplication of effort, have been largely avoided.

A foolproof interface that can be used by absolutely anyone should be enough to make this a priority area for computer manufacturers, especially as its impact on developing countries could be of culture-changing dimensions. And as, on the face of it, three-dimensional interfaces can increase the amount of computer storage remarkably, we would expect to see the Western computer industry taking a cue from the Japanese, and going hell for leather for this technology. It may be – or it may not. Large computer company executives play their cards so close to their body that one suspects they are glued to their chest wigs. Recently, Xerox dusted down an old toy, Rooms and produced a three-dimensional, flat-screen system called the Information Visualiser. IBM have admitted to a Virtual Reality research section. It is believed DEC also have one, and Siemens are certainly working on VR storage and retrieval. Quite what Olivetti, Amdahl, ICL or Thomson are doing is a mystery, never mind the Apples, Suns and Compaqs. The odds are they are doing something; it seems inconceivable they are ignoring developments.

Yet there appears to be a reluctance among many computer professionals to take Virtual Reality seriously. To be sure, most are serious, sober people devoted to running their companies, concerned by recessions and preoccupied by cash flows, invoices and the iniquities of suppliers. They have little time for a system which produces 'games'. And this applies even more strongly to computer workers in the large installations. VR appeals to the creative rather than the corporate. This indifference, for cultural rather than financial reasons, could be the nemesis of the European computer industry. It is one thing to watch from the sidelines, then move in when the time is ripe; it is quite another to ignore the game in its entirety, especially when the Japanese competition has its scouts out in force.

Virtual Chapter 6

It was to be expected that engineers would be the first profession to pursue practical applications for VR. Their training, and concentration on problem-solving, always leads them towards promising new solutions. Aerospace engineers are currently the heaviest users of VR. Over the last twelve months Boeing and BAe have been joined by Fokker in Holland, Northrop and Rolls-Royce in either using or developing VR-based systems. Led by Roy Kawalsky, BAe have continued to experiment with head-up displays, HMDs, tactile feedback and, above all, human factor problems, mainly in their military division. Fokker have been working with Virtual Presence on virtual robot arms, Northrop on design details and Rolls-Royce have commissioned interactive simulations of their new Trent engine.

The Trent work, done by Bob Stone and Andy Connell, was partly responsible for the creation of the new industrial consortium, VRS. For UK, indeed European VR, this is a very significant event. Not only does it include many of Europe's leading companies, so involving them in practical VR, it also ensures a funding level which will enable the best available equipment to be used – and developed. With hard work and a fair wind, VRS will drive the applications side of VR, with consequent spin-offs in the basic technology. Its work is varied. Stone is engaged in sub-sea visualisation, collision detection, environmental studies and, with Connell, nanopresence. (Nanomanipulation is on the agenda at UNC too.) Stone is also evaluating proposals for NYNEX, working with that organisation's CAD people.

In Germany the Fraunhofer Institute plays a similar role to VRS in the UK. It has opened a second demonstration centre in Darmstadt, where Martin Gobel's work on man-machine interfaces and converting CAD data into VR environments complements the consultancy, industrial robotic and training work in Stuttgart. The space side of aerospace continues to build on its VR expertise. NASA has an exciting virtual wind-tunnel project as well as a telepresence underwater robot (operated in Antarctica from California with a one-second lag). This is being used as a stalking horse for a telepresence Mars Rover. Paradoxically, an artificially large time lag has to be introduced to simulate problems accurately. The European Space Agency (ESA) has followed NASA and now has several lines of VR evaluation research, including telepresence, astronaut training, satellite control and design validation. It modelled a virtual Columbus space station (users handle a spaceball in their left hand, a data-glove on their right), and is producing the Andre Datasuit.

Aerial photography plays an important part in identifying geological deposits, water, indeed ancient civilisations, but interpretation is difficult. One British company has proposed using VR visualisation techniques to make the results more widely understandable. But satellite photography provides an altogether different dimension. A US military demonstration unveiled Project 285 to make a virtual reproduction of the entire planet by automatically rendering satellite photographs into graphics. Even this pales into insignificance next to the American project which plans to make a virtual map of the known universe, based on radio-telescopy. It requires two Cray Supercomputers, has been funded and started.

Back on the ground, VR is starting to make a limited impact on transport. Commercially used driving and car simulators are now in operation in Japan, the US, the UK, Spain and Germany. Dutch State Railways have commissioned a virtual stretch of track on which they are experimenting with the user-friendliness (or otherwise) of new signal systems, and a German haulage company is using VR to help it handle and route hazardous substances. Dassault Electronic have demonstrated a remotely driven van which they call DARDS. The driver, sitting in a control station some miles away from the test vehicle, uses a real steering wheel, pedal and gears to operate the van using a three camera video-link. While this is a tele-operation at present, there is no theoretical reason why the same system could not be used with telepresence for night or fog remote driving.

At least one commercial VR financial system is up and running in the US. Maxus Systems International have developed Metaphor Mixer, to track stock prices. It is being used to manage the $105 billion portfolio of TIAA-CREF, a college teachers' pension fund, and apparently has been excellent in helping to spot trends in Pacific rim companies. The grid displays up to ten thousand companies in one world, and can update at a staggeringly fast twenty frames per second. Each company price is represented by an icon. Price charges are shown by height and/or colour (red down, blue up, grey unchanged). The icons also blink or spin when fulfilling certain pre-programmed conditions, for example becoming ex-dividend or reaching a particular P/E ratio. Other features such as industry or country trends are also shown with colour changes. The visualisation solves the problem of making sense of a mass of rapidly changing data, and an interactive feature allows a full investigation of any chosen stock.

British Telecom (BT) has a very different problem. It has a massive complex network (six thousand exchanges and twenty-five million customer lines), which needs managing, and is evaluating VR to determine if it can help do so. Indeed, its overall VR research at Martlesham Heath had developed into probably the most advanced in the UK. Its main thrust is to visualise its network. Professor Peter Cochrane, Head of BT's Systems Research Division says, 'We must visualise in order to conceptualise, and we must conceptualise to get engineering solutions.' On a less philosophical track Paul Rea, a senior BT researcher claims, 'Three-dimensional representations have the advantage of reducing

113

the visual complexity and ambiguities associated with the more traditional two-dimensional forms.'

BT uses two methods. In the first the user navigates around the network, in the other the user gets a 'God's-eye view'. In either case, users can 'drill-down' into specific parts of the network to take a more detailed look. The view at present is that it is a very promising avenue. But BT is also exploring other avenues. It has developed a headset allowing field operators to be linked to a central control. Because the headset incorporates a camera, microphone, light and laser pointer as well as micro-TV screen and earphones, experts at control see and hear the same things as the field operative. Control can then advise verbally or by placing diagrams on the operator's screen. BT sees markets for this idea in security, law enforcement, medical emergencies and news gathering. It has also combined with Sheffield University to design a low-cost positional tracking system. Using a weak magnetic field it is being tested at Ipswich General Hospital for endoscopy tracking without X-rays. Longer term, BT is exploring VR-based 'what-if?' scenarios and the generation of VR models from data-borne descriptions – especially for its network management system.

Networks will be vital to business use and two are now being seen as the possible bases for distributing VR between users. Computer Supported Communication Work (CSCW) is a form of E-Mail for computers which allows several people to collaborate on the same task. Up until now Multiple User Dimensions, also known as MUDs, have been used by computer users to get together to play whimsical fantasy games. However, they can be used in VR. Xerox is by no means alone in its demonstrations of 'white-board' use or graph editing where several computer users work on the same project simultaneously. With the Phillipe Queau-led French expertise in televirtuality, it will not be long before virtual world conferences take place on a commercial basis. And this is before the widespread use of the fibre-optic super-highways. By the time they arrive VR may well be a proven technology, ready for take-off.

7

Education, Training, Health and War

Educationalists are great optimists. They are for ever seeking their Holy Grail; a tool which, at one and the same time, will make teachers' lives easier while better presenting a picture of the world to their pupils. There have been several pretenders, radio, film, television and computer labs among them; but each foundered, either because the equipment was not readily available, teachers disliked it, or it could not deliver what was promised. But, in truth, only a quality change was on offer anyway; a different medium for teaching subjects the same way as before, but more efficiently. Television is but a prettier, wittier blackboard, and computers (and perhaps multimedia) modern versions of the 'times table'. But Virtual Reality is truly different. It may well signal the end of the quest.

All too often education and training get confused, yet they are very different. Education widens horizons, training fines down on to a single technique; indeed a good education makes the subsequent training easier. As of now, educational Virtual Reality is only in its conceptual and development stages. An early experiment (1990) in the use of VR at the Navato Unified School District in the San Francisco area has now been followed by one pilot scheme with school-children in Britain, and at least two in America, where children at the public schools in Seattle and at the private Nueva Learning Centre in Hillsborough, California have been creating their own virtual worlds.

As a research scientist at the HITLab, with teaching qualifications, Meredith Bricken was the ideal co-ordinator of the first major pilot study into the use of classroom Virtual Reality. The state of Washington is committed to educational high tech; it has instigated plans to link its schools with fibre-optic cable, partly so they can network VR. It also runs a summer camp, the Technology Academy for both children and teachers at Seattle's Pacific Science Center. It was here in the summer of 1991 that five groups of children aged nine to fifteen took part in the classroom tests. With technical help from HITLab, they designed and created virtual worlds of their own. 'The children co-operated and collaborated,' says Bricken. 'We thought it would take longer for them to get into the idea, but we all underestimated their ability. The longest it took a group was three days.'

Preliminary evaluation has been positive. Children found VR compelling, and were motivated sufficiently to complete the project; indirectly they learned programming, networking and design. Teachers found it exciting, and liked it because one world engaged seven children at a time; it was unintimidating; they could retrace steps; it had a short learning curve and few instructions; and it kept them, and the children, on their toes.

An overwhelming volume of teacher and student anecdotal evidence exists to suggest that boys outperform girls on computers, so it is encouraging to hear Bricken. 'Girls loved it ... the process, rather than goal-oriented system ... they were more creative, more whimsical. They produced non-violent worlds, no blowing-up of anything.' Most children grasped the idea intuitively. 'I would like to use VR to go back in time to see the making of history...' 'I would like to go into a computer as an animal ... to see what it eats,' wrote two of them. But what they are more likely to get is Algebra, now in development by Bricken and her husband William. This transforms a conceptual world into solid models, where children can play among the equations. For the large numbers of children who cannot imagine what algebra is about, this colourful, noisy, three-dimensional world is a perfect way to learn and, crucially, to understand.

VR is not only active, as opposed to the passive world of television and books, it accepts imaginative inputs from students. For example, a virtual world has no gravity, magnetic field, echoes or reflections. Children can learn physics by programming *different* physical constants: say, a ball falling upwards; and a teacher, networking into the world, draws the appropriate lesson. In theory, anything can be experienced. Be a zinc sulphate molecule in a reaction, a plant in a drought, or even a snowflake. And VR is natural. As Meredith Bricken puts it, 'The skills needed to function in a virtual world are the same skills we've been practising in the physical world since birth.' It is very much learning by doing and experiencing, rather than just accepting the wisdom of others. Bricken makes a crucial point when she points out that in Virtual Reality, 'What you *do* is what you get ... not just what you see.'

Michael Clarke, head of West Denton High School for thirteen to eighteen-year-olds, on the edge of Newcastle, decided to do something himself. He developed the idea of Intelligent City. Sixth formers are creating a virtual city where students will be able to shop in virtually real French and German streets, discuss their orders in the appropriate language, and calculate and pay in virtual francs or marks. In effect this is the world's first VR language lab. The school has funding from the Department of Employment, matched by private funds, all to the tune of almost a hundred thousand pounds. The 'city' is only one of their ideas. The school has also built a virtual factory for a Health and Safety education project – do the wrong things and you trip, or get bricks on your head; and the third project uses VR as a medium in which users can envisage the best sites for works of art in public places. This also has an impact on the art itself.

The main users are fifth and sixth formers in the arts and technology strands and, as Bricken discovered with her students, they enjoy creating the worlds at least as much as operating within them. But the students themselves find the system helps enormously; one told us, 'My GCSE project's to design a housing estate. It's one thing to put it on paper, but it doesn't half help when you can actually walk around it.' Another lad said, 'If I'd built a model of this street it would have taken two or three months, but it only took me two weeks to do this ... and it's better.' The school uses desktop VR to keep costs low, but plans to get a projection screen and hopefully an immersive head-mounted display if it can raise the money.

West Denton is the first European school to use Virtual Reality, and is building on its large existing stock of computer experience and applications. Clarke is an enthusiast. He believes VR is an ideal teaching technology as it fits school patterns perfectly. 'It can be used by large classes or small groups; students can keep projects in their own files, and it crosses disciplines and subjects,' he says. 'The software is robust enough to be used for anything we can throw at it. We can create libraries of worlds, and use them for the school as a whole.' Whether the money will be there for him to carry through his concept is another matter. In Britain, at least, education is well down the list of priorities, behind the armed forces, health, unemployment pay and sundry emergencies.

This money shortfall, the need for both better cognitive and technical research, and making VR pupil-proof, all mean it will be some time before virtual worlds invade most classrooms. Nevertheless, VR appears to be made for education. Because it is easy to use, entertaining and visually stimulating, it also offers more to children and adults with learning difficulties than any other technology. The University of Nottingham's Department of Engineering and Operations Management has a VR consultancy unit which has been working in this area with Shepherd School. Young people with learning difficulties are being offered a familiar VR learning environment in which they can manipulate symbols of everyday objects with virtual hands.

Virtual Reality will provide a new, interesting educational environment – and once you capture the interest of students, the rest is relatively easy. VR appears to have the facility of keeping children quiet by absorbing their attention fully, and at the same time stimulating unexpected flights of imagination. It is the nearest we have yet come to general, classroom-based, learning by doing. Teachers who have tried it are enthusiastic, and those who have only read about it are intrigued. Even Russian educationalists are investigating its potential. In the long run, it is doubtful if it could be kept out of schools, even if that is what we wanted.

Mandala is the Canadian VR system which uses video-cameras to include users in a virtual world. Users see themselves on a screen in this world – like a mirror, except that it is *not* a mirror image – and interact with the virtual objects in it. It has been used in the home, in clubs, public places and schools. In Ontario schools it has been described as a 'step into Sesame Street' for young children, and as 'a challenging, teaching video game' for older ones. While Mandala is used to teach dance, music and rhythm, it also has academic uses, such as language teaching, basic reading skills, business charting, and economic data presentation. Although it has been received well by students it has a drawback. Users must keep their eyes fixed on the screen in front of them, as they would for television or ordinary computers, only even more so to cope with the interactivity. In time Mandala, which is now available in America, Germany, Britain and France, will be networked to form the basis of distance learning.

As the 'motor' behind modern simulators VR (including Mandala) will be capable of touching most physical jobs. At present it is used by pilots, by train-drivers, by fire-fighters in Orlando, Florida, and it saved the American College of Aeronautics from a dreadful fate, a forced move from New York to New Jersey. The college had been polluting the atmosphere with fumes from welding practice, and was about to be exiled. Along came Ixion, a Seattle-based company, to develop a virtual welding trainer, with a virtual welding torch, colour and heat changes, and bad welds.

Before the end of this century Virtual Reality will be used

for training in at least some, and perhaps most, of the following areas: pre-road car driving, ships' masters and pilot courses, control desk work in nuclear or continuous process plants, health and safety procedures, engineering skills (tool-setting etc), deep-sea diving, oil-rig work, plumbing and other construction jobs, car mechanics and other servicing work, trade skills (e.g. electricians), space duties, police riot control, disaster responses, first aid, fire-fighting, complex knitting, and a host of medical, dental and veterinary techniques. And this list is only a start. But whatever the training subject, VR lends itself to an imaginative approach. William Bricken has suggested a training program for assembling a carburettor, where the pieces squeak if put in the wrong place – presumably they will purr if housed correctly.

But its uses go further today, and will go even further tomorrow. Mandala is being used to improve musical skills, manual dexterity, reactions and balance. Both the French and Americans have produced virtual musical instrument packages, the French in Grenoble with force feedback, and both can be used for instrument teaching and practice. Indeed the Americans Rory Stuart and John Thomas have suggested worldwide linked virtual master-classes, suitable for the entire range of concert instruments. But why stop at music, why not theatre, art, physics or history? VR also will be used to train people for public speaking, and for dealing with members of the public. It will be available to train the planners and staffs of events like big receptions, tours, elections, sales, and pop concerts. Virtual Reality will be a trainer's dream: cheap, flexible, easy to use and evaluate. All we need is a continuation of today's steady technological improvements.

If education is all about tomorrow, medicine is about ensuring most of us get to see it. To help in this objective, the training of health workers has become one of the prime objectives of Virtual Reality research. Virtual cadavers may sound macabre, but in the not too distant future, they may be the normal way for medical students to learn anatomy.

For some years now, Suzanne Weghorst at the HITLab and the Department of Biological Structure at the University of Washington Medical School, jointly have been building a high-resolution digitised cadaver. When complete, students will be able to wield a virtual scalpel, in precisely the same way as they do a real one – but without the mess. Anatomy can be difficult to learn from two-dimensional books: but VR will take it from the dissection room to the coffee table. And when functional graphic models are incorporated, of, say, circulation and nutritional absorption, students will be able to interact with these virtual bodies to learn about physiological, chemical and pathological changes. Dr Frankenstein was clearly ahead of his time; he should have waited for Virtual Reality.

But it does not have to be the whole body. Dr Joe Rosen of Dartmouth Medical School in America is working closely with VPL, developing Virtual Reality for medical uses. They are designing a virtual leg, with all its joints, and the forty-one muscles. The idea, however, is not to dissect it, but to examine how it works, so that reconstructive surgery (especially for sports injuries) can be put on a more rational footing. Rosen, a plastic surgeon, has also been working at reconstructing external parts of the body – a form of virtual clay modelling – prior to operations. Indeed, pre-operative planning is one of the major potential uses of VR.

Existing medical technology is starting to merge with Virtual Reality. Rosen and VPL are in the process of producing a generic virtual colon. Although partly for training, it is intended also to link with both the diagnostic MRI data and CAT scans. If these were to be superimposed over the generic colon, there would be a three-dimensional picture which would enable a surgeon to practise on a customised 'living' virtual version of the patient. Virtual Reality is already in use: at UNC doctors have been inside a virtual thorax to position beams of radiation precisely at a real malignant tumour, and X-ray pictures are being 'enhanced'. With a helmet, extra computing power, and more accurate mapping procedures (all to allow CAT, MRI and X rays to be translated directly into 3-D graphics), doctors will be able to walk around a tumour to determine its extent, and so form an accurate prognosis and treatment plan. As part of this research, the HITLab is collecting data from real cadavers, so as to prepare a reference database which distinguishes precisely between healthy and diseased tissue. This system should be up and running within ten years.

Medical training does not stop at qualification. Virtual cadavers and patients will be used as 'test-beds' for the manual skills of surgeons prior to their being awarded their Fellowship, and as practice facilities for rarely performed procedures, especially in heart and brain surgery. Dentists could test surgical extractions or nerve blocks, while Virtual Reality would seem to open up multipurpose and open-ended horizons in veterinary surgery, especially for training. However, excellent force feedback is essential; without it practitioners will have lost a vital tool. So we are talking in terms of ten years rather than months before such surgical techniques can be used on live patients.

Ixion, the welding simulator company, also produces medical training devices. Its chief executive lightheartedly refers to them as 'flight simulators for physicians'. One system allows doctors to practise endoscopies without damaging patients in the process – it is estimated that a doctor needs to perform fifty endoscopies before becoming proficient. The training doctor puts the endoscope into the mouth of a life-sized model head and chest. If the doctor moves in the wrong direction and hits the model oesophageal or stomach walls, he or she encounters a force feedback sensation that feels the same as the resistance encountered by a wrong move on a live patient. The path of the endoscope is tracked by video on an adjacent screen. This kit is so popular among American doctors that Ixion is now America's third largest producer of endoscopes. Ixion also produce a simulator for practising keyhole gall-bladder procedures using a mixture of video – for the organs and endoscope – and graphics for the insertion of the catheter into the bile-duct. The company also produces a life-size doll (reminiscent of those sold in less salubrious establishments), on which resuscitation techniques can be practised. Using force feedback, it is connected to a computer which lets you know, verbally, and with computer screen life signs, how successful you are – the doll can actually die! None of these needs helmets or special gloves. They are the equivalent of the two-dimensional screen version of VR.

Life has a habit of copying art, even if it can take a long time to catch up. Many years ago a movie, *The Incredible Journey*, was based on the idea of shrinking people so they could travel through the bloodstream of an important patient. In theory, a telepresence operation will be able to be mounted within a patient, the operator being in the position of a blood corpuscle, ferreting through the body's highways and byways, viewing and scanning as it goes. Action will then be taken, either by administering the appropriate medicines, micro-surgery or the use of micro-telepresence robots which will open arteries and ducts, hoover away cholesterol deposits or snip off rogue cell formations. However, this will be some way in the future, even though most of the basic technology is in place today.

The future of health will be inextricably intertwined with Virtual Reality. As improved medical techniques and drugs create an ageing twenty-first- and twenty-second-century population, so there will be an increasing demand for sophisticated medical services. Virtual Reality will help spread limited resources more thinly. Today Bob Stone at the ARRC is liaising with British surgeons to develop VR techniques for ophthalmic and keyhole surgery, and with Glaxo to visualise what migraine sufferers 'see'. This surgery research, and that being done in America, will ultimately lead to telepresence-based keyhole surgery – but good force feedback will be needed – with or without robotics. Within the next half-century conventional invasive operations will be as redundant as the old top-hatted, butcher's apron, pre-anaesthetic procedures are today.

There will be many more medical uses. Among them will be audio/visual analogues of patient's vital signs in operating theatres and intensive care suites, both of which are in development now. Telepresence medicine could also be practised in remote or underdoctored parts of the world – the Australian outback, and parts of Africa, Asia and South America, or in space. Virtuconferencing and round the clock monitoring will allow severe problems to be dealt with at home or in family doctors' surgeries rather than in hospitals. Body computers will check health continuously, automatically sending warnings to doctors the moment blood chemistry or pressure start to show signs of pathological change. Electronic templates of the movements and posture of a fit young person will be replayed into that person's system when they stoop and creak with age. The dream of eternal youth will be edging just that little bit nearer. But there will be a downside. If continuous health monitoring is used on those in the prime of life, it will make VR the world's most effective creator of hypochondria.

Quacks will exploit this. They know people believe in computers and will peddle VR nostrums from their Harley Street broom-cupboards, of as little use as the old carpet-baggers' tonics, but with the cachet of being made available through the 'infallibility' of computing. We have already seen similar 'scams' with lasers, often with dreadful results.

But medicine is not only technique. Healing may be more about the will to get better. Today, some cancer sufferers mobilise their inner strength by visualising their defence cells attacking the tumour, willing them to succeed, and visualising the tumour shrinking under the assault. Using Virtual Reality you can be a virtual lymphocyte, and aim the attack yourself. You will be able to touch the cancer, punch it, pull bits off it – even hear it scream. Or you may be a cell fighting a chronic infection, or the AIDS virus. VR can act as a healing agent in so many ways, helping to calm the spirit, to find the inner self, and to reinforce the will to fight and live. It will be like living the dream of visualisation, the dream of life and health, rather than death.

Dave Warner is visualising something completely different. Working at the Loma Linda University Medical Centre in California, he is using VR techniques to visualise and understand the invisible causes of neurological disorders. He describes his work as 'perceptual psychophysics', studying the correlation between brain activity and physical and sensory actions. He uses a DataGlove to record precisely the tremors of Parkinsonian patients, which are then correlated with treatments, and he uses the Convolvotron to test brain activity after constant sounds. Warner is developing what he calls the 'microscope of psychiatry', trying to forge new ways to approach neurological accidents, degenerative and congenital disabilities.

Although inaccurate, the assertion that many severely disabled people live their lives in a Virtual Reality is highly evocative. But VR has the potential for doing massive amounts of good in this area of life, especially the use of Virtual Reality gloves. These include the Teletact II, the Pisa Kevlar tendon glove, and the Rutgers glove with pistons. The British glove is intended for use by patients, the other two mainly for rehabilitation work, monitoring progress after operations or accidents. The Teletact was tried on a woman with zero hand mobility at a Los Angeles conference in early 1992. Although it took some time to calibrate it properly, it finally gave her a considerable amount of mobility – it can work in practice, not only in theory.

Walter Greenleaf, the founder of Greenleaf Medical Systems, has the exclusive medical-purpose licence on VPL equipment. He uses the DataGlove for his Glove-Talker. This ingenious invention allows a person with speech difficulties (post-stroke or laryngeal trouble) to communicate by means of text on a screen and a voice synthesiser, by gesturing with the DataGlove. The gesture vocabulary is customised for each patient, so maximising the use of existing abilities. First reactions from patients are highly favourable.

Greenleaf's company also makes a Gesture Control System, enabling a severely disabled person to control devices, and a Motion Analysis System which measures movement. This latter idea can be used to play back correct nerve impulses and muscle movements to people who are

125

having to relearn them – accident or stroke victims for example. Bio-Control Systems has invented and produced a different product, Bio-Muse, which is used by badly damaged people, quadriplegics and the like. It turns tiny eye, lip, swallowing or other muscle movements (using elasticised bands) into music or textual communications, and allows the user to move virtual objects on a screen. And even now there is a demand from restricted mobility, wheelchair- or bed-bound people to experience complete freedom of movement in a virtual world. And it can feel very real. A young wheelchair-bound boy playing virtual football with a Mandala system got so involved he fell out of his chair. Blind people have been guided around rooms by the sounds from a binaural system, and it is possible to start rehabilitation procedures with VR – for example juggling slowly with virtual balls, and then speeding them up. However, it is too early to get carried away. The systems are all in development stages, and depend on charities, altruism by companies like VPL, or venture capital for funding. Only the rich or the very fortunate will have access to Virtual Reality-based relief in the immediate future.

Illness involving the mind is every bit as real as physical sickness. In theory, VR should be able to address the entire range of psychological illnesses. The ARRC has started research into arousal with the University of Edinburgh, and is collaborating with the University of Leeds to use a VR system to handle phobias such as arachnophobia by using 'virtual spiders'. Given its singular chameleon-like, or even wish-fulfilment properties, virtual worlds would appear to have a role in the treatment and management of simple, and perhaps social, phobias in general, not to mention sex and aversion therapies. Indeed there will be some who will wish to use Virtual Reality as an escape route for their clinically depressed or psychotic patients. However, there is an obverse side of the coin. Virtual Reality may actually cause or exacerbate some of the types of problems it is intended to alleviate. There are also a number of very difficult moral and ethical dilemmas, all of which will have to be resolved, not to mention the dangers of misuse in the wrong hands.

If psychiatry is fraught with moral and ethical problems, so is the defence industry. After all, the main function of most weapons is to deter by being able to kill better than the enemy. Because the search for more efficient, powerful weapons goes on all the time it is no accident that so many technologies get their first development and test runs in the defence industries, especially in America. And, despite recent stringencies, it is still one of the few areas where money is, relatively, not a problem. The early work on virtual military cockpits is spinning off to virtual simulators. A conventional simulator is looked at head-on; what happens to the side and behind is unknown. But in a virtual simulator pilots can see friendly, or unfriendly, planes on either side, and if they turn around, behind them. It was on these virtual simulators that many Gulf pilots did their training.

The Gulf War has been proclaimed the first 'Virtual War', because pilots had practised over and again on three-dimensional computer graphic terrains built up from satellite imaging. The pilots probably knew their routes over Iraq and the Saudi desert better than the route their children took to school. Indeed, they may well have been superimposed on their helmet visors, especially on night flights. However, virtual cockpits are another matter. Because the modern warplane is a complicated web of computers and other electronic gadgetry, it is hugely expensive. And as it gets more complex, so it gets more difficult to fly; even on non-combat missions. This is tantamount to putting the large investment in the plane (and the pilot's training) at risk because of foreseeable human errors. Virtual cockpit programmes are designed to relieve pilots of many of their visual, and possibly decision-making, duties.

Some descriptions of virtual cockpits concentrate purely on head-up displays (three-dimensional images on the pilot's visor). However research has gone much further than this. It was realised early on that peripheral vision was as important to a pilot's well-being as head-on, so this had to be incorporated. At the American Wright Patterson Air Force base they are trying to augment the sense of sight with enhanced aural and tactile stimuli, as well as presenting information in graphical form. Flying by touch, by sound, but blind to the real world is not out of the question. In Britain the work at BAe, and its contractors, developed a visual display system which is generated behind the pilot and fed on to 'spectacles' using mirrors. The company is also concentrating on how graphics can be made more usable in practice.

While it is impossible to determine precisely how far this has reached in defence circles, it is acknowledged that much of this is cognitive research which tends to be subsumed under the umbrella title of 'human factors'. It leads us to ask an important question. If we cannot reproduce the real world – because as yet computers are not sufficiently powerful and viewing screens not yet sufficiently detailed – what level of picture would be acceptable? And what is the minimum detail pilots need, so they know what they need to know? Until we get these answers and take the necessary practical steps to implement them, the virtual cockpit will not be good enough for combat use. It is one thing being able to represent zones of danger on a graphic map by wire-frame domes, it is quite another making them accurate enough for a pilot's life to depend on them. This does not mean that work has stopped on the technical side. Research into better head- and hand-tracking devices, non-distorting wide-angle optics, faster computers and better software are all still priorities; and as we have seen, deep inroads are being made into these problems.

If nothing else, the Gulf War demonstrated that the modern world is not the most perfect of places. It raised the spectre of nasty forms of warfare. For example, one good reason for wishing to equip pilots with VR in a virtual cockpit is to protect them against blinding nuclear flashes. This is not a humanitarian consideration, it means they will still be able to fire off bombs and missiles. And if it is not war, it is often terrorism. At Brooks US Air Force base, where there is a virtual airport with runways, hangars and flying planes for use in air traffic control training, they are also using Virtual Reality for anti-terrorist, SWAT Group deployment training.

At the highest level, warfare in the late twentieth century is conducted by one technological marvel piled on top of another. And Virtual Reality is taking its place among the marvels. Infantryman 2000 is an army concept, developed at the University of North Carolina. A soldier is insulated from chemical, bacterial and nuclear attack by an all-enveloping suit, and sees virtual images on a helmet display. And from people to the impersonal. Virtual Reality was used as a 'market-research and design' tool to adjust the button sizes on Patriot Missile instrument panels – they were too big!

Training can be expensive. General Electric markets a three-dimensional product called Compu-scene. At an advanced level it involves simulated tank, helicopter and warplane flights and battles. It has excellent picture quality, but can cost considerable amounts of money – up to thirty million dollars for one helicopter simulator package, and the American army has over one hundred and fifty helicopter simulators. So if a technology comes along which can reduce the skill levels needed to work with it (and so overall costs), it will always be considered carefully. At present radar and ASDIC rely on the interpretations of trained operators. However, if the data is presented in a real-time graphic virtual form, anyone can 'be' where the planes, ships or submarines are – and determine what they are. Costs are reduced, and in emergency more people can be drafted into the work. Marconi produce a similar range of simulators for the UK armed forces, including cockpits, radar crews and periscope work, and currently are evaluating three-hundred-and-sixty-degree head-mounted displays.

But some virtual training schemes not only cut costs, they may be the only schemes available. With worldwide disarmament gathering pace, and environmental objections (against using the countryside for tanks, or low-flying warplanes), the physical space for military exercises is becoming restricted. For example, it is likely that the only way to rehearse, and refine, NATO battle plans in the near future will be in a virtual world – similar to the General Electric model. And if the battle-plan involves areas such as the Antarctic, jungles, deserts, rain-forests or even the Himalayas, it is clearly far cheaper to simulate these terrains than ship troops and equipment to them. In any event, think of the immense cost savings in capital equipment, fuel, maintenance, manpower and accidental casualties, wherever the training has to occur.

SIMNET is a worldwide Virtual Reality, real-time, simulated tank battle game. It has over two hundred and fifty networked simulators in Europe and America and a number of simulated 'real' terrains. It also involves helicopters, planes and buildings, and obviously is considerably cheaper to run than 'real' manoeuvres in Germany or wherever. As with shared W Industries video games, all the other tanks in the game are manned by simulator crews. The 'tank' interiors are realistic, the army crews 'fight' and enjoy the training, as do the logistical staffs, who get to practise their jobs. All, that is, except medical personnel, burial teams and chaplains. The American navy has now developed a comparable training system Battle Fleet In-Port Training (BFIT), while the American air-force is not far behind. Ultimately all three will be integrated into NATO training systems.

The soubriquet 'war games' has never been more aptly applied, except perhaps for a game developed at the US Naval Post-graduate School, called NPSNET. This can run up to five hundred land-, sea- or air-based vehicles simultaneously, all of which can be user controlled, from networked work stations. However, in addition to helicopters, ships, planes and tanks, it uses heavily armoured hamburgers, and a heavy-lift aerial bovine vehicle (a flying cow)! Bizarre it may be, but nothing like as strange as many arcade games are – and Virtual Reality games will be.

Virtual Chapter 7

A worldwide economic recession is not the best of times to launch radical projects, especially if they need government funding; and the educational uses of VR come squarely into that category. As a result they have marked time over the past year. Certainly some existing projects have gone from strength to strength, notably the VIRART teaching of educationally disadvantaged young people in Nottingham's Shepherd School, using Makaton symbols. Michael Clark, who introduced VR into the West Denton High School curriculum, has transferred to south London, where he may manage to sow another VR seed. Before he left, however, he used his considerable powers to persuade the United Nations Standing Committee on Trade and Development to use West Denton School as a test-bed for a virtual trade-point. As this is the first of its kind, it is a definite step towards the establishment's acceptance of Virtual Reality as a viable concept.

Training is in a similar state. While car- and train-driving simulators are now being used or developed, a ship's pilot training scheme is under discussion and Hughes Redifusion's aircraft simulators become ever more powerful, sophisticated and expensive, most commercial companies are fighting shy of even modest start-up costs. It would appear that for VR training really to take off the mass-entertainment industry will have to force down the price of the basic kit, or there could be a synthesis of education and entertainment ('edutainment') for matters such as driving simulators. There are, of course, some developments. The mixing of VR with multimedia is a logical step, and Perry Huber from Faberushi Design has demonstrated a 'virtual exhibition' where you can move between virtual stands; and on each stand a virtual video has informative real multimedia displays. Meanwhile the space agencies continue to use VR as a training medium. One of these new systems catches the eye. Dr R. Bowen Loftin and his team at the Johnson Space Centre have developed a virtual laboratory, complete with gravity sphere and floating control panel through which physics students (or trainee astronauts) can control gravity, friction and time. In such an environment abstract concepts such as gravity, magnetism or relativity become realities.

Medical training is potentially a heavy user of VR. Although that potential is far from being fulfilled, increased amounts of research and development are being carried out, especially in the US. In California a crude but anatomically correct virtual abdomen – through which a user can 'fly' – has been designed by Dr Richard Satava as a first step on the path to a complete surgical trainer.

But there has been considerably more movement on the application side, especially in techniques helping disabled people. Some of these are worth noting and following. Dr Luigi Tesio from Milan is testing a system of immersive VR for ataxia sufferers – people with a very poor sense of balance. Many of them have serious problems, but their vision tends to compensate for the loss of other functions. Tesio subjects patients to different types and intensities of visual clues in a virtual world, so isolating the real problem. Patients can then learn to 'listen' to their own bodies. Tesio admits, however, that a proper feedback system needs to be developed for his system to be really useful.

The Swedish Handicapped Institute is using robots with VR interfaces. They are, however, pre-programmed to perform certain limited tasks. But BioMuse from BioControl Systems is anything but pre-programmed. Although available a year ago, in the intervening period it has been trialled, tested and refined. Its reliable neural interface (Enabler), now makes it an all-purpose computer interface, as capable of driving a telepresence robot as of eye-tracking or making music. It works without muscle movements, relying on external bands detecting the nerve impulses which would have made the muscles work, had the person not been paralysed. It may well have an extensive range of other medical uses, from the treatment of eye disorders to body mapping, and the monitoring of stress, muscle and nerve tone. The next phase of research, which reads like something from *Star Trek*, is to convert thoughts of muscle movement into an interface, and perhaps even emotions.

Another practical use is in Barrier Free design. Wheelchair VR is a wheelchair placed on fixed rollers, with friction pads to simulate slopes. The operator wears an HMD, a glove and uses a joy-stick. It was designed to give architects the proper illusion of wheelchair hazards and so take appropriate action. In the US it is illegal to construct new buildings that are not barrier-free. The first public demonstration of Wheelchair VR was actually inside a Chicago flat designed with its help. The wheelchair can also be used as a convenient method of locomotion through a variety of different virtual worlds.

There have been other moves in the past year. Gloves in Italy and the US have been used to measure the degree of repetitive strain injury, and in one case analysed the hand movements of the Boston Red Sox star baseball pitcher, Roger Clemons. And Massimo Bergamasco in Pisa is leading a multinational team developing a robust glove interface with force feedback. There has also been a successful attempt to superimpose a graphic VR image of an ultrasound scan on a patient's abdomen. Other work going on in Italy includes using VR for micro-surgery in eyes, general micro-suturing and micro-force-feedback.

Colonel Richard Satava is a US Army surgeon, whose ambition to be an astronaut was thwarted at the quarter-final stage. Now he has settled for being the first person to perform an operation in space, even though his feet will be firmly on Californian ground. Satava has been experimenting with telepresence surgery on offal, using robotic surgical instruments designed by SRI International. It is a telepresence extension of the newish technique of minimally invasive surgery, also known in the UK as 'keyhole surgery', and in the US as 'Nintendo surgery'. He sits at his surgical work station using force feedback surgical manipulators on tissue which he sees graphically on an HMD, while the surgical instruments are following his hand movements in another part of the laboratory. Telepresence surgery would, for example, enable a consultant surgeon in London to assist in operations in Alaska or Zaire, without ever leaving Harley Street. (Howard Paul has used similar robotic instruments to replace the hips of arthritic Alsatian and Retriever dogs in Davis, California, with a startling ninety-two per cent fit of metal shaft to bone, compared with the normal twenty-seven per cent of the average orthopaedic surgeon.)

Savata freely admits that better graphics, proper force feedback which can simulate the different elasticities of our internal organs, and more powerful computers are needed, yet is convinced that telepresence surgery is feasible. So enthusiastic and persuasive is he that the US military has agreed to have a VR surgery input into its 1995 exercises.

While much of the military work with VR is shrouded in secrecy there has been some information about the simulation side. Now that 'Star Wars' has been abandoned it is probably the most prestigious high-tech defence research in the USA. The current military expenditure on graphic simulation is a not inconsiderable three billion dollars per year, and the Pentagon anticipates throwing around three hundred billion dollars at computers and electronics over the next ten years. For example, Florida company IS&T has a six-million-dollar commission from the US Army Research Institute just to build a 'World-class VR research lab' to find ways to use VR as a training tool. And the US Army has asked IBM to head an impressive consortium to develop a new shared virtual environment training system. Initially built around M1 and M2 tanks and vehicles it will have nearly five hundred modules. With the planned integration of air force, engineers and artillery, the cost will be well over one billion dollars.

However, this may be a substantial underestimate. The US Army alone has an annual training budget of $2.3 billion, and as Thomas Edwards the Deputy Chief of Staff in charge of its training has said, 'Training is simulation'. He also suggested at a recent conference that VR would underpin future Army training techniques because it had the capability of keeping up with this rapidly changing world. (He neglected to point out the cost-savings.) In broad terms he could see VR being used in battle terrain learning, vehicle and ordnance use and maintenance, personal and group tactical awareness and plain old combat training.

Much of this Army and other US Forces funding will go towards creating the next generation of new VR equipment. For example a proposed pilot 'debriefing system' using virtual environments will be partly developed in the Sarnoff Californian laboratories. One of the reasons for this is that Sarnoff already works on many of VR's enabling technologies, including displays, advanced computers, interactivity and intelligent agents. The debriefing system (which will also be used for pre-battle briefings) will concentrate on how the pilots' performances affect their colleagues. In other words group co-ordination and co-operation rather than the usual individual training.

There has already been a virtual re-creation of a Gulf War tank-battle, the Battle of 73 Easting, fought out on a digitised map of the Kuwaiti desert. This SIMNET simulation was so accurate that during its development it was discovered that different tanks claimed to be on the same piece of ground at the same time, while pilots claimed to have shot Iraqi vehicles that had already been destroyed. In the heat of battle memory becomes unreliable. But there is also a very serious point. At the time of the actual battle the US forces knew the geography of the desert better than the Iraqis, indeed the Kuwaitis themselves. Technology beat local knowledge.

But the most ambitious project is a seamless battle simulation system, involving all the US armed forces, which will make Simnet look like Pong, the earliest computer game. The Institute of Defence Analysis (IDA) suggests it will have ten thousand separate battle station simulators representing submarines as well as tanks, and fighter-bombers as well as sappers. The terrain would be utterly accurate, mapped from satellite transmissions – and could be anywhere in the world. But the important part is the seamlessness. It means the division between reality and simulation will be deliberately blurred. In modern warfare much of the combatants' time is spent watching bits of the action on screens and periscopes. But it is now possible to put any images (real or imagined) on them. Players in these new war games will not be able to distinguish between real events on false screens or false events on real screens. That should keep everyone on their toes!

This is not a game, however. This is big money; indeed vast amounts of money. And yet it is so much cheaper than using conventional training and exercises. But this is not the only area of savings. The US military now realises that simulations prior to manufacture can save both billions of dollars and time. Virtual factories building virtual planes and tanks, then the real thing – a seamless design and production process. And with the new US rule that military research must have civilian spin-offs, simulation and VR will be well placed as prime candidates for considerable amounts of funding.

8

Entertainment and the Arts

It is a surprisingly long step from computer *war* games to war *games*. The idea of the former is to provide a realistic, detailed battleground on which both new and old battle techniques and tactics can be practised. In the latter, the backdrop can be relatively primitive, and the game itself becomes the most important thing. But there are more differences than this. In a *war* game the non-participants in, say, helicopters or buildings must have an inbuilt intelligence (or be capable of being controlled); at the least they must act as friend or foe. The realism must be at such a high level that all support staffs must feel able to join in without feeling supernumerary. This requires considerable computer power. But the war *game* player stays within a limited number of options and outcomes, and far less computer power is needed, so keeping costs down. As this applies to all games, not just battles, it was inevitable that games would become the first commercially viable Virtual Reality product to reach the public.

In March 1992 the first major feature movie to be built around Virtual Reality was released in America. *Lawnmower Man*, based on a Stephen King short story, beat a rumoured five others (including Gibson's *Neuromancer*) to the screen. It was to have been accompanied by America's first VR arcade game, based on the movie's leading character, Cyber-Jobe. Unlike the W Industries games, launched a full year ahead of it, it did not have a head mounted display. This is not because of any new and better technology, rather that American health and safety laws frown on them. The possibility of catching head-lice and the inability of players to hear or see a car entering arcades were two of the reasons for banning their use. The more plausible excuse of being unable to detect a potential mugger was not put forward, perhaps because arcade owners who argued for the ban did not want that particular possibility publicised. The opposition from owners was based on the fear that their VR equipment, gloves and helmets, would be vandalised or stolen, while entertainment industry rivals were preaching the gospel that wearing helmets would be like 'putting an old gym-shoe over your face'. It appears strange that according to American 'experts' European and Japanese VR games players disregard their personal hygiene, and have lost their sense of smell, as well as their aggression.

As a substitute the American game used a form of periscope, with the forehead pressed close against a headpiece. While this did not give three-hundred-and-sixty-degree vision it gave good three-dimensional images, and at different times these appeared to be behind the screen, on it, passing through, and then hovering between the screen and player. The interactive element came from a gloveless player's hand which went around and behind the 'stem' of the machine, and into view. But the helmet was not the only difference between a W Industries game and the American model.

CyberJobe was played against a background of video sequences taken from the movie. This relieved the computer of the task of creating real-time graphic backgrounds, so almost all its power was channelled into the graphics and game itself. This was lengthy, involved several stages, and ran like a narrative. Players went through various levels, trying to intercept invaders against a background of different scenarios. In effect, it was the world's first Virtual Reality, interactive, immersive movielet.

However, because of a mega-miscalculation, only the prototype was ever produced. The investors felt that with no stars *Lawnmower Man* was not a big enough Hollywood film to sustain a subsidiary game, and arcade owners had just seen a hyped-up, high-tech arcade game – albeit not Virtual Reality-based – fail. But *Lawnmower Man* reached number two in the American movie charts, and stayed there for some time (it's comforting to know the experts get it wrong too).

Although the game itself was innovative and good to play, the real technological interest lies in the part of CyberJobe. He was played by the same actor in the movie and in the arcade game. But he did things in the game he did not do in the film and, moreover, he did not act in any additional scenes. Impossible, but true. The actor was body-mapped. This means the game directors took his body and, using one, or a combination of different techniques, made it do what they wanted, where they wanted, and when they wanted. Body-mapping is a simple technique where a subject is 'photographed' from all conceivable angles, using not film, but lasers. The resulting electrical impulses are converted into a computer program, cleaned up, and played back as a reconstituted body.

The resulting image is an exact electronic representation of the performer, not a graphic. In principle it is not unlike the still photographs which are 'wired' by telephone. Although it is possible to map an old performer, for example Marilyn Monroe, she will be in two dimensions – proper body-mapping is three-dimensional. The body-map is brought to life either by other actors wearing a type of DataSuit or body jewellery, or by computer animation techniques. Greg Panos has a scheme to body-map all well-known and up-and-coming performers in a specially designed mapping suite. Performers will be mapped on a treadmill, move, laugh, cry and generally emote. But there are artistic and ethical matters to be considered. The essence of a performer might well be lost. In other words the body-map may look just like Robert Redford, but the 'puppeteer' wearing the DataSuit (or the animator) will most probably not be able to act like him. And who has the copyright?

Panos intends performers and their families to keep it, but it is unlikely that all producers will be so ethical. How will Equity, the performers' trade union, react? And how will the use of body-maps be controlled, especially if they are used in criminal or immoral acts? Body-mapping is fraught with difficulties which, as it is here to stay, will become ever more important.

Games are an important element of modern life, both from the standpoint of a healthy person and a healthy economy. Games allow people to play, relax and enjoy themselves; it is a mistake to think of them as trivial. They even reflect the cultures in which they are played; for example, a British arcade game has to be 'juiced up' with extra blood, violence and screams for Japan, but in Germany (where they levy a 'blood tax') the deaths and violence have to be removed, while in Sweden the same game can feature the deaths, but with neither blood nor sound effects.

Virtual Reality is made for games. It creates the perfect playground of another world with infinite possibilities. Where W Industries has blazed the trail, others will follow, graphics will improve, as happened in ordinary computer games, the flicker will lessen and the time lags diminish. Within two or three years, immersion will be that much greater, probably exceeding Jonathan Waldern's ninety per cent target.

There will always be a demand for a games arcade. It acts as both a social and an entertainment centre, especially in smaller towns. Over the years the nature of these arcades has changed. The old mechanical pinball and fruit machines have been replaced by a wide variety of electronic games, and although many of the attractions are of the video viewer type, with hydraulic movement, taking you on rides down ski-slopes or up cable cars, Virtual Reality is beginning to make itself felt.

The most recent incarnation of the arcade is the Simulation Centre. In it you can play virtual pool, where you use a proper cue to start, or one of the W Industries' games. Using the helmet (Visette) and a joy-stick you can be in a battle, fly a Harrier or drive a racing car against four other players in Total Destruction. Legend Quest, which opened in Nottingham in February 1992, became the world's first Virtual Reality theme park. Four people in their own stalls, equipped with Visettes and joy-sticks group together in a specially designed and decorated theme room to play a fantasy game on W Industries equipment. In the fantasy section, players may choose from a range of identities, such as wizards, dwarfs, or elves. Their voices are electronically altered and together they fight skeletons, spiders or whatever in a co-operative, three-dimensional interactive form of dungeons and dragons. Each player sees and hears the virtual world through the Visette, which has also been developed to include tiny microphones, enabling players to communicate with each other in the course of the game, so adding to the team effect.

These sort of games with more than one player at a time, in the same shared space, have proved to be very popular. Arthur Wickson, owner of the Rock Garden, an old-established rock club and restaurant in central London, was the first person to see their commercial attraction – and instal one. 'No regrets at all,' he says. 'It's been a success ... there's a constant demand to use it day and night ... and some people keep coming back to play it. We're looking at opening up in other places.' And playing is not cheap: three pounds for a three-minute slot. Reactions differ. Some people complain of dizziness, nausea or palpitations and it is alleged that one user suffered such severe vertigo after being picked up by a pterodactyl that he had to obtain medical treatment. But this does not deter people from coming back for more. 'Best bloody thing in London,' said one twenty-year-old. 'It's the challenge, getting the other guy,' was the view of a regular player in the club. 'Well good ... rad ... well good,' was the view of another young man who had just spent fifteen pounds on the game. A very different set of players frequent the world's first VR nightclub. Cyberseed, which is also in Covent Garden, opened in late April 1992, and looks towards the 'concerned cyberpunk' market.

Open from 7p.m. to 3a.m. three nights a week it appeals to people who are as intrigued by all the possibilities of the technology itself as much as by the games – which include both W Industries machines and Mandala.

W Industries is now taking its games to America. The company has signed an agreement with Horizon Entertainment, which is owned by Edison Brothers, a firm which also owns over two thousand shopping malls, stores and 'family entertainment centres'. Clearly Horizon has a different clientele from the conventional arcade. The company is not unduly worried about vandalism, or about a lack of hygiene. In any event the Visette comes equipped with removable plastic strips and wipes. Given the success of Battletech in Chicago the W Industries–Horizon partnership might be a long and lucrative one.

Battletech is an ambitious, real-time, fully interactive game which is played on crowded simulators. Set in a politically unstable thirty-first century, it is the civilian equivalent of SIMNET (except the four-player tank simulators only play against one other at a time), and it appears to be played with an equal ferocity, and devotion, by its young acolytes. The game itself is so complex that a lengthy briefing is given to all new players in an on-site 'briefing room'.

Virtual Reality also allows the possibility of customising arcade games. Why not have break-dancing, or a driving game, set within graphic representations and videos of local streets? Good idea? Too late. An American company is trying to sell it at this moment. And it is inconceivable that Las Vegas casino owners will ignore VR's potential as an interactive gambling medium – virtual blackjack or roulette perhaps. And having recognised it, the next step would be to put it on networks. Overall, there is little doubt that VR will make arcade games bigger, better and more sophisticated, following the example of their theme park cousins.

The recent changes in theme parks are a harbinger of the future. Where once the most popular rides were big dippers of gut-wrenching exhilaration, log flumes with horrendous drops and wet finales, and white knuckles were *de rigueur*, nowadays the longest queues are for a simulated roller-coaster, or space ride, most often based on a blockbusting movie. New proposals by AGE to build the world's first VR amusement park, Perception Circus, outside Osaka in Japan, will go beyond this. Its roller-coaster will use 'sim-cabs' suspended on hydraulics, with passengers wearing helmets, and both sound and wind will give the impression of being on a real ride. The feelings might be the same as in the old parks, and the surroundings are certainly more magical – but the element of danger, of risk, however slight it might have been, has gone. Truly, disbelief will have been suspended.

But there are plans to take VR theme parks even further. Discussions about creating a virtual zoo in Leicester are well advanced. You not only see the animals, you find out about their lifestyles, life-cycles and habitats as well, using VR and multimedia. And it is suggested that AGE would like to create a virtual aquarium in Osaka, where viewers could participate by swimming among the fishes. A virtual aquarium already exists in America: at Marriott's Waterworld viewers look through wide-angle optic portholes into a virtual tank. The subjects for VR entertainment applications are legion. Why not behead a few virtual aristocrats in a French Revolution theme park, play golf on the moon (the short holes are four miles long), act Othello, or drive in the Monaco Grand Prix? And why are we substituting the virtual for the real? Partly because as we will not have to travel, pay entry fees, or buy special food or clothes, it will be cheaper; and partly because, as we shall see, world cultural values are pushing us in this direction.

But if theme park games are neither real time, interactive nor immersive (although immersion is very high in rides like Back to the Future), games with these qualities are about to hit another market. Home-based Virtual Reality games are just around the corner: W Industries aims to have its system on the market within two years, and Sega and Nintendo would like to beat that. When 'home reality engines' reach affordable prices, the genre will take off. Even now, non-VR disc-based systems such as Amiga's CDTV are selling, yet they are far from cheap, while people are creating their own games with the Dimension 3-D toolkit. Home immersive systems will be equipped with all forms of interactive and optical devices; gloves, wands, space-balls and hand-grips will be sold alongside three-dimensional goggles. However, systems which play only VR games will not sell well, and will probably have to accept ordinary adventure games as well. And home VR 'doodling', constructing one's own worlds, and playing or hiding in them, will please a different age group and slice of the population.

Whether the arcade-style W Industries games or the more whimsical Mandala-style products get the lion's share of sales is problematical. Indeed they may not actually compete directly. Mandala is less of a game and more of a 'fun' machine, and it is being used in clubs as much as in homes for activities such as virtual body-painting (the electronic version of the 1960s technique for rolling paint-covered bodies over a canvas), virtual music playing, and pop video productions. And while Mandala had the advantage of being on the market first, that may count for nothing in the longer run. Indeed, the overwhelming percentage of users surveyed at the 1992 VideoArts exhibition preferred wearing a helmet to using the large screen. But overall, if the 1980s boom in personal computers and games software is any guide to the future, there is little doubt that fortunes will be made in VR, and well before this century is out. Certainly the big guns, like AGE and Nintendo, are in the market, although this may be the one area that leaves room for the pioneers, or the smaller entrepreneurs. Fibre-optic cabling, when it arrives, will add another ingredient, although this is one for the big players, the telecommunication and television companies.

Mainstream television companies are notoriously conservative, as indeed they need to be to target their central core of viewers. But cable and specialist companies are proving to be more adventurous. Interactive television is currently being tried in America. It started with sportscasts where viewers were given a home 'videotron' switch, and could decide on camera angles and replays; now it has moved on to old shows. Series such as *The Lucy Show*, are being cut into small bites. Viewers have to decide what ending they want – and use their switch to make it happen; they can also play games with them. The interactive technology is working, and although it is most definitely not home-VR, it is a step in that direction. People are learning that they can interact, and companies are finding out what they like interacting with, and how they do it.

Two other factors are stimulating this market. The partial deregulation of American telecommunication companies (telcos) has given them the opportunity of participating in providing material for the cable networks, and this they are eager to do, especially with interactive, educational programmes. The telcos plan to use fibre optics for the trunk traffic, and high-compression ordinary cable from the kerb. The huge computer companies are providing the other stimulus. They have just woken up to the fact that only thirty per cent of American households have a computer. But if they place one in every television set, their production would rocket. This is what they are aiming at. And what would computers do in the set? They would allow for games, for interactivity – and, if powerful enough, provide a 'home reality engine' for Virtual Reality programs. And there are proposals for digital television, leading up to the potentially interactive High Definition Television (HDTV). Even Paramount, the very mainstream movie and television production studios, has commissioned a report on interactive entertainment.

Virtual Reality itself is about to star on television. While there have been programmes on VR – the BBC's *Colonising Cyberspace* was particularly imaginative – none of them actually integrated the technology. Mandala features in a show called *NickArcade* on the North American Nickelodeon Cable Channel, where children compete for prizes in a virtual world. A specific interactive VR game is planned for showing on British national television by BBC 2 in October

1992. The game, Cyberzone, was designed by, and runs on software produced by Dimension International, the British company which sells the least expensive of the home 'toolkits'. Two teams of two players (one a 'warrior', the other a 'guide') try to out-think each other by using VR alter egos ('puppets') in a mythical city, where a 'controller' makes life more complicated for them. It is intended to use twelve huge immersive screens for the players and floor pressure pads and clothing sensors to detect the warrior's movements – which are then translated to the puppet. The game is run on five ICL PCs, two each for the teams, and one for the interfering controller. Whether it becomes popular is in the lap of the gods, but because viewers cannot interact (at least until HDTV comes along) it may need the addition of a game show element to succeed.

Interactive movies or theatres are a reversal of the historic trend; they are having to catch up with interactive television. And much of the stimulus is coming from Japan – in the shape of Hollywood. Jaron Lanier of VPL and MCA (owned by Matsushita) are jointly developing a form of interactive theatre, the 'voomies'. In outline form the experience appears to be more like a do-it-yourself game than a normal theatrical experience. There is an audience of about thirty, six of whom sit at computers and play the role that the controller plays in Cyberzone; that is, they can aid or distract the players. They all play in a virtual world, which apparently includes a free-spirit 'Puck-type' figure, called a 'changeling'. He or she will be a combination of dancer, stand-up comic and compère. This interactive performer will have to 'bring out' the shy, calm the over-exuberant and, most importantly, bring the show to a close. The audience is 'included' in the action either by wearing their own helmets or by using a huge screen. As it is still in the developmental stage this plan may change before its public debut, which will be in both Japan and America.

The fact the Japanese are driving this form of entertainment demonstrates that their commitment to Virtual Reality stretches beyond videophones, robotics and visualisations.

In any event the Japanese are avid users of computer games, possibly because of a lack of space in their homes and the high cost of building entertainment centres on expensive land. With Matsushita owning Universal, and Sony another large Hollywood studio, Japanese companies are in a position to fuse their technological expertise with the stock of old films and television programmes they have 'inherited'. This will steer both interactive TV and new forms of cinema and games, and one would expect movies produced by both companies to spawn associated VR games. Furthermore, as Sony and Matsushita also have interests in theme parks, it has to be assumed that commercial synergy will prevail, and that an increasing number of their attractions and rides will be based on VR.

Yet all sorts of non-technological problems need to be overcome; for example, a new set of conventions and cues will need to be laid down – those established for the movies are inadequate. It would not be appropriate to use the sort of music which we all know indicates danger or happiness, or to have villains wearing dark hats. And without conventions, explanations will have to be made, and a lot of subtlety will be lost.

With so many new tricks to learn, a new generation of film-makers is descending on Hollywood. They are equipped with visions of new cinema forms, and their heads are filled with the techniques and technologies of computers, graphics, animation, optics and VR. Basically they are more technologist than artist, and their conversations revolve around mips and polygons rather than blondes and grips. One thing is certain, the arts of animation and graphic designing will explode as VR takes hold in the entertainment industry. Their movies will be a combination of the mechanical (hydraulics), binaural sound, three-dimensional screens, tactile sensations and smell. Interestingly, one of the very old school of film-makers, Morton Heilig of Sensorama fame, is still talking of producing Sensorama Simulators and an Experience Theatre which, although completely immersive would not be interactive.

But if Heilig is not producing real VR experiences, neither will the new film-makers. Virtual Reality is experienced from a first person perspective; it is inclusive and it is interactive. Every member of the audience will have to be given their own helmet, and allowed to interact with the production. This might be prohibitively expensive. Clearly, interaction with film is limited at the best of times – it has to be sliced up like the old TV programmes. And with a screen, even if the film was entirely animated, how would the audience decide what it wanted to see? By voting? Can you imagine losing, and watching the 'wrong' ending?

Nicole Senger, working from the HITLab, has overcome these problems by creating a VR film intended for an audience of one at a time. *Angels* gives an original perspective on paradise, where each graphic character has a life of its own. Interaction with the movie and its characters is by hand gestures. Senger has used symbolism to reduce the number of polygons needed, and viewers claim that the resulting lack of realism is more than compensated for by the intensity of the experience. The user can move anywhere, and even see the dirt at the back of the set! With a solitary person as the audience, *Angels* is nearer to a work of art than a movie. And as a working proposition it probably will find a home in art galleries rather than theatres.

In its way, art creates as much controversy as politics or religion. Some people believe everything must be understood, if not actually liked, by everyone. Others appear to believe that art can only be true if it is understood by the cognoscenti, and them alone. So most people would not agree about what constitutes a good piece of sculpture; but most would agree that sculpture should have a physical form. But not in Virtual Reality. The use of two fingers and two thumbs can create exquisite three-dimensional virtual representations. It is sculpture. It looks solid. It can be handled. And once the boundaries of objects have been fixed, and force feedback (or tactile simulation) has been improved, the object will feel similar to conventional, solid sculptures. And if the artist is wearing a DataGlove linked

148

to a Puma-type robot, the VR sculpture can be replicated as a physical, wooden sculpture at the same time.

It is possible to paint in VR, get performance art in VR, or use it as part of other performances or exhibitions. Mandala is one way to do this. Music, dance, painting, or even making people disappear, just like the transporters in *Star Trek* is part and parcel of its stock-in-trade. Interactive art exhibitions are starting, although for the most part they appear to be rudimentary and obscure. In San Francisco there are plans to open an interactive art gallery which may, or may not, survive. There have been public performances of dance and theatre using virtual props and backdrops, and performances by artists wearing helmets and DataSuits, whose 'puppets' also feature in the performance. Beverly Reiser has even written 'Life on a Slice', a Mandala-playing, interactive, multimedia poem.

Institutions are springing up all over the world. In Britain, the Virtual Reality User's Group (VRUG) whose aim is to explore all VR avenues and, in the words of its co-ordinator Martin Kavanagh, 'democratise and demystify Virtual Reality', takes two-dimensional screen demonstrations around clubs and gigs. Although it intends to pursue its aims through the worldwide bulletin board network, it also meets regularly. At one of its meetings the Finnish artist Pekka Tolonen, the writer of the Symbolic Composer computer program which converts physical entities into music – for example the proteins in the AIDS viruses – gave a paper on LULU. LULU is a black and white semantic model which runs on an Atari computer, has a will of her own, and responds to written instructions of an explicitly sexual nature – it was censored during its public debut at the Finnish Science Centre. Although produced for two-dimensional interacting he is converting it to run three-dimensionally on the appropriate equipment, as will its male equivalent program Don Juan. VRUG also pioneered video-conferencing when they 'met' the French Virtualistes group in a virtual space somewhere between Camden in London and the IMAGINA conference in Monte Carlo. Kavanagh says, 'This shows the great potential ... but we've got to be aware of the great dangers as well. Meanwhile, the group will carry on encouraging, and helping producers or anyone interested in the subject.'

The German government part-funds the Institute for Image Media in Karlsruhe. Geoffrey Shaw, the institute director, is a media artist whose work, The Legible City, is attracting international interest. A viewer of this world uses an amended exercise cycle to move through a city composed of letters and words, seen on a large screen in a darkened room. The city is based on the real city plans of Manhattan and Amsterdam. The Manhattan texts are from people with an intimate connection with the city, among them ex-Mayor Koch and a cab-driver. Amsterdam is different. The letters are all in different styles, conforming to the shapes of the buildings they replace. Viewers are free to take any route they wish.

But the Germans are not alone in acknowledging VR art. A Dutch piece by Agnes Hegedus and Geoffrey Shaw is based on a three-dimensional fruit machine. Users try to assemble a row of fruit in virtual space to win a virtual shower of coins. The Canadians funded a six-week seminar in Banff in the Rockies on Virtual Reality in the movies, looking at VR as a subject, a technique and a medium. Three new VR pieces emerged, one of which, Silent Room by Kathleen Rogers (who gives seminars on VR to art colleges and at the British Film Institute), has been prepared for television and another, a virtual American Indian 'longhouse' by Lawrence Paul, may find a place in the Vancouver Museum. The ICA in London has run two totally crowded lectures on Virtual Reality in a series on the new Media Arts, and before long will be obliged to feature VR art in one of its galleries. Exhibitions and forums, from Cyber Arts International to Siggraph, show what artists can do; masterclasses are given and expertise swapped. YLEM is one such group, and others are springing up like mushrooms in stables, all based on the new media. Clearly VR is a medium ready and waiting for the expressive artist to exploit it; more permanent than performance arts, yet neither solid nor immortal.

Where, then, does this leave Virtual Reality? Unfortunately there is a tendency for some people to jump in with both feet, only to find someone has drained the virtual pool. The San Francisco Art Institute ran a series of VR events. As reported in *CyberEdge* they were amazingly well attended, yet the speakers knew little or nothing about VR, or indeed about art! The discussions centred on war, drugs, sex – and VR's role in delaying the news. On the other hand a great deal of energy and creativity is being expended to bring the benefits of VR to people. This involves both the uplifting of the spirit and of the body. The battle to make sure that Virtual Reality will be used for precisely these reasons is only just beginning.

Virtual Chapter 8

It has been at least as good a year for theme park rides as it has been for *war* games. New rides seem to appear each month, zippier, faster, more realistic, more frightening and more exhilarating than the others. But as wonderful as they are, they are not interactive – at least not yet. But now large corporations, from Sony to Paramount, are developing rides with varying degrees of user interactivity. Driving games, laser-tag games, interactive *Star Trek* rides and games, and all sorts of simulators will soon be delighting children, both young and old. Last year virtual theme parks were promised in Japan and the USA, but neither materialised. The latest promise, from Iwerks, is that they will open Cinetropolis – including interactive VR experiences – some time in 1994. We shall see. Now even Europe is jumping on this particular bandwagon with Brian Eno and Myron Kreuger involved in a virtual theme-park project in Barcelona.

Although many of today's rides and games are labelled Virtual Reality, in reality there are very few, and those that are genuine VR are in arcades or special centres. Most are doing very well. Battletech has outgrown Chicago, is expanding in Japan, and was sold recently for fifteen million dollars. W. Industry's games are played worldwide, while its Legend Quest remains the best themed VR. It is also technologically innovative with 'smart keys' containing club members' particulars, character and game preferences and playing records, while also deducting the time spent on games – like a phone card. It is an encouraging sign that research has shown that not the least of the attractions of these centres is the social interaction, both in the real and the virtual worlds.

VR Slingshot is really the first home-based, affordable VR game. Produced by Ixion, which makes endoscopy simulators, it is played on an Amiga, and allows players to compete in the same virtual space over a telephone line or direct cable link. It can be either 2- or 3-D, the depth coming from 60hz 3-D glasses. Also for home consumption is the electronic quilt, a worldwide co-operative art fantasy. Interactive TV has also continued its steady progress. BT is doodling around with the idea of TV cameras with fish-eye lenses, home HMDs and a viewer zoom facility. Another pay-TV method in the US includes a personal choice of camera shots and replays. And while Cyberzone, the world's first networked VR TV game has finished its run, it is likely to be reconstituted in a similar form.

The past year has seen more than a little ingenuity in presenting graphic images. The Cave, presented at the shrine of computer graphics, SIGGRAPH '92, was advertised as walk-in VR. Cube-shaped with displays in front, on both sides and another on the floor, it produces a different type of immersion. A 'driver', who changes the scenes with a wand and tracker attached to the electronic glasses, is 'elected' from among the six viewers. Although brilliantly conceived, the lags were too long, and there was a difficulty correlating changes in scenes to the actions of the 'driver'. At the same exhibition Sun Microsystems showed the Portal. This lift-sized cabin 'descended' and immersed its two 'travellers' in a virtual aquarium. Domes are also fashionable, both for flight simulators and at SIGGRAPH as a visualisation tool. Real VR was represented with both Myron Kreuger's Tiny Dancer where a participant was shrunk and placed in the giant hands of the other. The same sort of effect was created by the new Mandala.

During the past year it has become clearer that it is not just the technology holding back VR entertainment systems, there is also a lack of commonly accepted metaphors, instantly recognisable icons – and good scripts. While this applies to all facets of VR, the market reacts far more sharply to adverse publicity in the entertainment industry. So it is ironic in the extreme that VR is proving to be surprisingly effective at helping to create storyboards and sets for ordinary feature films, as well as being featured in them. Indeed, the latest BBC News studio uses a seriously virtual set.

Much to the disappointment of the media, virtual sex has remained barren over the past year. There was indeed a virtual orgy, but it used virtual sound technology only. The best of the rest was a rehash of Virtual Valerie, winsome enough in her own right, but not the stuff of the future. Cybersex seems to be on the same time-frame as a cure for AIDS. But body-mapping, 'artificial life' and 'virtual puppetry' did take off, all of them prerequisites for virtual pornography, among other things. And among these other things are marketing and publicity.

Artificial life is about inanimate objects being given a biological (and intelligent) behaviour pattern. Also known as intelligent agents, they can be anything, from brilliant blobs to clever carrots. They populate virtual worlds in both games and educational programs. While much work is being done in France at the MediaLab, the main protagonist is Fujitsu whose algorithms drive these new entities. But if these interactive creatures are in the background, then virtual puppetry is in the foreground. Eggwardo made a highly acclaimed appearance in the Loma Linda Medical Centre Children's Hospital. A genial sort of cartoon character, he joked, chatted and comforted the patients. They saw him on a TV screen and talked over the phone. But this technique, known as

Performance Animation System (PAS), has now gone a stage further.

Tarbo is a friendly, interactive dinosaur invented by Steve Glenn's Simgraphics. Whether Tarbo-chan (a Japanese dinosaur) has been welcoming guests to a hotel or talking to them at conferences, she (or is it he?) is played by an actor with special head, hand and partial body equipment which tracks movements and transmits them to the graphic dinosaur. A similar Simgraphics system called VActor Performer was played at SIGGRAPH, and involved a virtual clown projected on a viewing screen who joked with, and directly questioned, startled passers-by. And a PAS-controlled Mario has promoted Nintendo to a host of adoring youngsters.

Away from the home, exhibitions, theme parks and marketing, Cyberseed, the Covent Garden VR night-club run by Brian Davies which closed after nine months, has been revived in central London in a more structured form. It now provides a wider range of activities, including E-Mail, VR tutorials and discussion groups, Silicon Graphics machines on tap and a more settled format. That the demand is there is certain, the problem is to fit together the disparate VR techies, cyberpunks, hackers, techno-freaks and general clubbers. One of the original frequent visitors, an art student, concocted the world's first S & M VR event, a claustrophobic experience of virtual touching arms and threatening bodies.

VR art is not setting the world alight. It is true that some professional sculptors are fascinated by the possibilities of working in virtual space. One pointed out it would be possible to create a 'solid' object in a space that has multiple vanishing points and perspectives. This could enable sculptors to work within a medium that gives viewers the illusion of many dimensions, and where the boundary of the space is always uncertain. It would be like wandering through a hall of mirrors, inset into a maze. Geoffrey Shaw and his collaborators in Germany and Holland are keeping the flame burning, as are a variety of American artists, but it is really a set of familiar faces. It may well be that high costs are keeping new artists at bay at present, and that sponsorship is needed. But Martin Sjardijn, a Dutch artist, is using VR practically. He wants to put a mirror into space orbit and has been using VR to simulate both the technique needed and the results that would be obtained. And Art+Com has tried out a method of modifying a painting by just looking at it, partly using an infra-red based eyetracking device, which if successful would have dramatic repercussions on all VR techniques.

The Networked Virtual Art Museum is being created under the auspices of Carl Loeffler at the Carnegie Mellon Museum. There has already been a link between Munich and Pittsburgh (on the same lines as Philippe Queau's Paris/Monte Carlo televirtuality hook-up). The similarity continues as one of the virtual galleries is based in Munich where it is called the Fun House. In it users can borrow bodies, or clones, just as they do for the Abbey of Cluny. With virtual agents giving users a helping hand, a mirror that shows the back of the person looking at it, a virtual Da Vinci flying machine and various Egyptian temples, there's more than enough to keep users happy. Ultimately Loeffler sees the museum as an education resource transcending the five universities with which it is linked.

On a similar tack the Austrian town of Linz, which already hosts architectural and other gatherings which include VR, is to open an 'Ars Electronica Centre'. It is intended to become one of Europe's leading VR venues, blending the arts, science, research and education. Opening in 1996 to celebrate 1000 years of Austrian nationhood, it will be exceptionally well-equipped with five separate VR systems as well as hundreds of other computers, servers and electronic aides.

9

The Essence of Virtual Reality

Painters, sculptors, movie, television and theatre directors
all know full well the power of illusion. They use it to tell
their stories, to entertain and above all to challenge mun-
dane perceptions. And in the public eye illusions have the
reputation of being different and special; so much so that
up-market magicians call themselves illusionists. It is in this
respect that theatre and cinema, television and art have one
important thing in common with Virtual Reality: they
replace, and in the case of creative people, often deliberately
confuse, everyday reality with illusion.

Illusory virtual worlds are, however, different in certain
key respects from those created by the movies, television
or paintings. These differences are so crucial to understanding the potential of Virtual Reality over the longer term
that it is necessary to enlarge upon them. In essence the
differences can be summed up by VR's five 'i's: intensive,
interactive, immersive, illustrative and intuitive. Taken
together they forge a new, distinct medium.

And of equal, if not more, importance are the distinctions
between Virtual Reality and conventional computing:
indeed, it could be said that there is a world of difference
between the two. VR is to ordinary computing what live
theatre is to the movies. Live theatre is a vital, unpredictable
experience which varies with every performance, the performers reacting to their 'feel' of the audience. Indeed, in
some plays and musicals, performers and audience actually

interact in the same space, as they do in *Five Guys Named Moe*. On the other hand a movie has to be the ultimate in predictability; the audience knows exactly what to expect, no matter how often they watch it. In the same way, unless a system goes down, or a virus attacks, conventional computing is predictable. Indeed this is one of its greatest marketing points; users need to know exactly what will happen each time they use it. However what happens in a virtual world depends on the computer user. It is the user's inclusion in this illusory world, and the ability to influence what happens in it, that makes all the difference between Virtual Reality and ordinary computing, the movies, television or art.

Being inside a different three-dimensional world is a very strange and intense experience. But the sensory deprivation caused by wearing a helmet adds to this, so that the intensity created is far greater than in any other form of computer or electronic medium use. *Virtual Reality uses all your concentration, and fills all your senses*. If ordinary computing is like sitting in an armchair with a cup of tea watching a televised soccer match, then Virtual Reality is standing on the Highbury North Bank, being carried along physically and emotionally with the thousands of other Arsenal fans. And with this intensity comes enhanced impact and increased believability. The user is said to be 'immersed' in the virtual world.

Conventional computers are alien things. To use them we have to play to a strict set of rules, none of which is used outside computing. But Virtual Reality is an interactive medium. Users can make choices, and the virtual world responds. Furthermore the way the user instructs the computer is by a series of intuitive everyday commands, by voice, by touch, by gaze or by gesture. Navigation is by using hand-eye co-ordination and ordinary hand and head movements; the user chooses in which direction to move. It is possible to touch and feel objects, start and hear motors, and communicate with – and respond to – other people. This is not an alien world, this is an alternative virtual world, where things behave in much the same way as they do in ordinary life, or at times in dreams. Given the combination of the immersion and the 'normal' environment, the intensity can become overpowering.

Watching television is very much part of our everyday world. We can treat it as familiar wallpaper, ignoring or concentrating on it as we wish. It is a passive experience. But the interactivity, immersion and intensity make Virtual Reality an active experience. It is not an enhanced television service. Indeed as it gets into our homes there will be a similar quantum change in the psychological and cultural impact as was experienced when television replaced radio as *the* mass electronic medium. This change, and the sheer all-demanding nature of virtual worlds are important to bear in mind when trying to understand the uses and impacts of VR in 'brain-washing', politics, religion and marketing.

Different things appeal to different people. The wide range of available television programmes shows that clearly, as do the diverse types and large numbers of newspapers. These also make it very clear that different people prefer to get their information in different forms. In Britain, as in most countries, newspapers divide into tabloids and broadsheets, but the difference between them is less of size than of content and treatment. This is not only to do with the choice of subjects or photos. For example, the *Sun* and *The Times* may both run an article on the economy, but the only words common to both may well be the definite and indefinite articles. The fact that the *Sun* and other tabloids outsell the broadsheets several times over shows that the majority of people prefer to get their information in an outline, easily digested, popularly packaged form. At the same time it is also only too obvious that television programmes dealing with 'serious' political, social or economic subjects are relegated to very off-peak times. In Britain this has created a wide information gap in these areas.

Virtual Reality has a most valuable property in these circumstances. It is illustrative, capable of providing information in a visual and auditory analogue form, if necessary with a sense of touch added on. Where so many people fight shy of anything that looks like a graph or a statistical table, Virtual Reality can use readily identifiable analogues. Why bother with data when you can stand in a virtual supermarket, taking things off shelves, all to show how the economy is working, or be a jockey in a horse race demonstrating how political parties have been doing in the opinion polls. VR could make data accessible to those who, for a variety of reasons, have been cut off from it in the past.

The fact that sound analogues can be added enhances the sense of the visualisation. Indeed, research suggests that good sound analogues allow poor graphics to become believable. For a host of reasons, the visualisation of problems has not commanded universal respect; literate and numerate skills have been more socially, academically and professionally acceptable. But we depend on visual cues and information every day of our lives. There is no reason why detailed, technical information cannot be processed in this far more accessible form – for scientists and professionals, as well as 'tabloid readers'.

So Virtual Reality is intensely different. It absorbs (or immerses) users in a different, vivid world. It engages all our senses, all the time. It is relatively easy to use and can get its information across in a populist and accessible way. But on top of all this it is interactive. You, the user, will be part of the world. You will not see the racehorse/opinion poll analogue – you will be part of it. That is what will make it such a powerful tool.

There can be little doubt that Virtual Reality will play its part in the power games of the future, much as television does today. But we shall have to adapt to the changes that this will bring. Indeed, from health to education, torture to war and art to government, Virtual Reality will increasingly make its presence felt and change the way we perceive things. As time goes on, so the technology will progress, and move further away from its original aims, rather like gunpowder. Virtual Reality started in the abstract, and then moved on to war; perhaps ultimately it will bring peace.

So that we can take a look at these options, using what

we know about the technology today, the research, products and plans being hatched, we shall now look at the longer term. And as we explore the ways that VR will be used to save lives, or take them; speculate on bringing dead politicians back to life, or killing politics entirely, and consider whether we educate people in a new enlightenment or condemn them to a second Dark Ages, we must bear the five 'i's in mind. It is their combination that distinguishes Virtual Reality from television, movies and ordinary databases. It is their effect on people, for good or for ill, that will drive the changes.

Part Three

Glimpses of Heaven, Visions of Hell

10

Can There Be Another Reality?

On an intellectual level the concept of another reality has
been contemplated for centuries, and still continues to raise
insoluble questions. On an everyday level the idea of there
being more than one reality is generally not uppermost in
our minds as we struggle to survive in the only one we have
ever known. Although philosophers from classical Greece
onwards have agonised over the precise nature of reality,
and science fiction writers have speculated about other
worlds, never before have we been in a position actually to
enter a different reality. A century ago a group of British
philosophers hypothesised they had invented a 'reality
machine' which would enable them to enter other worlds.
They met to discuss, in the abstract, what impact such realities would have on individuals and society. Their con-
clusions were as unreal as their premises. But today, at last,
we have that machine, and we are being forced to consider
the practical effects.

Another reality is a very powerful idea. Over time it has
been used as a weapon of social control by disreputable
historians who unearthed imaginary glories of realities long
ago so as to foster nationalism. And futurologists and science
fiction writers have been accustomed to anticipate coming
realities, with a view to changing current behaviour. So both
past and future realities, the one appealing to nostalgia, the
other to progress, have exercised considerable power over
human thoughts and deeds. But never before have we had

163

two realities we can affect running concurrently. The potential power inherent in this schizoid situation, whether it is seen as a control mechanism, a money-spinning device or liberating experience, is simply gigantic.

All too often, definitions of new machines degenerate into collections of acronyms and jargon, understood only by the chosen few initiates. Others are lists of dry statements on how to connect mysterious pieces of equipment which make as little sense as your pension fund rule book; others just obscure enlightenment. For example when combining the definitions of 'virtual' and 'reality' in the *Shorter Oxford English Dictionary* we get: 'Something which is real, or has actual existence, but not formally or actually.' Precisely. Virtual Reality, however, has attracted a splendid definition from one of its gurus, John Perry Barlow, amongst other talents the lyricist of the Grateful Dead. Quite simply it states: 'Cyberspace [Virtual Reality] is where most of your money is, most of the time.' And this definition brings us back to the nature of reality in a way that everyone understands, because it affects our wallets and purses.

We can safely assume that most people hold their wealth in banks, building societies, DSS giros, pension fund accounts, or bonds and shares. Nowadays few people use a tea-caddy or mattress as a repository for their savings; and only a small proportion of income or wealth is held in the form of ready money, which we even call 'hard cash'. The rest exists in a vast global battery of computers, identifiable only as minute pulses of electrical energy. This is not money as we know it – try giving your children pocket money this way! In the worst-cast scenario a simple fuse, or a mischievous hacker, could wipe it all out without us knowing until we tried to get hold of it. Yet we believe this money exists. We believe it because we use the same system to pay bills by cheque or credit card, and apparently our creditors are satisfied. We can even draw hard cash from hole-in-the-wall cash dispensers, which proves that at least some money exists.

As in VR we can change the system (by inputting or

removing money), or the system may change our behaviour by telling us we have no money to make a purchase, or even, horror of horrors, eating the card before our very eyes. Yet money is real because a consensus exists that it is real. Hard cash promises to pay the bearer the amount printed on a bank note and if you present it to the Bank of England, demanding your rights, they will give you a similar note in return. But money itself, be it the 'hard cash' of a pound coin, a dollar bill or an electric pulse, has no intrinsic value. However, as it is in everyone's interest to believe in money, it has acquired a real-world value. Yet there are American hotels and gas stations where you cannot pay by cash, only by electrical impulses – partly for security reasons, partly because hard cash is no guide to credit-worthiness. Try renting a car in California without a credit or charge card. Credit and virtual money is preferred to real money.

Reality is what we believe to be real. Take the classic philosophical question. If we look out of a window and see the houses opposite we believe they are real, but without being over there to touch and walk around them, how can we be one hundred per cent certain? And when we are there, and have verified they are real, we look out of the window at our own house – and the same question arises. And when we go away, and cannot see any of them, can we be absolutely certain they still exist? Indeed how do we know far-away countries exist? And we know from Einstein that everything is relative: the houses may be growing at fifty feet per minute, but as long as we are growing at the same rate we shall never notice. Reality is relative. However, a consensus exists that the houses are real. The neighbours, visitors, the postman and milkman, indeed the people who live in those houses believe they are real; and so they are, just like money.

But is what we see actually reality? Not according to the biologist Richard Dawkins. He has suggested that the way people understand and find their way around the world is not by a direct reference to the world outside – but to a virtual facsimile we create in our brain. This virtual model is updated only when the brain is faced with a perceptible difference between the world outside and the virtual facsimile. Perhaps this is why the memories of the streets of our childhood remain locked into the view taken by a child. We are often shocked when we return to see how small it really is – we had this model of a large virtual world locked into our memory. If Dawkins is correct then when we are immersed in a virtual world, we produce virtual facsimiles of virtual images.

Virtual Reality is just another reality. We can move around and take an active part in it; and in turn it can change our behaviour. The fact that it is computer-generated, with no physical existence, makes it no less real to the people in it at the time. It is not unique. Most of us create other worlds in dreams or daydreams, and these are real enough at the time. And people hallucinating on drug-induced 'trips' or schizophrenics listening to their auditory hallucinations, know their experience is only too real, while other people have intense 'real' religious visions. On the other hand, some Amazonian Indian tribes have never experienced money; to them it is just useless pieces of paper or metal. They do not believe in it. In the same way no one has to believe in Virtual Reality until they have experienced it.

We have created and observed other realities from Stone Age times; Egyptian tomb paintings, Renaissance sculptures, photographs, movies or television, all represent different realities. Some were scenes from everyday life: bison, trees and people. Others, like the statues of Amon-Ra, Hieronymus Bosch's hell, the brooding city of *Blade Runner* or Dr Who's Daleks have stretched our imaginations into new, unreal realities. Computers play a latter-day part by creating alien worlds, inhospitable dungeons and monster-ridden mazes in ubiquitous arcade games. And now the Holodeck in *Star Trek 2* and films like *Lawnmower Man* are presenting us with images of Virtual Reality itself.

This is the crucial point: they are only presented to us; we only observe them. We are not in them or of them; we cannot play parts in them and can only affect the outcomes to a very limited degree, and then only if a 'Dungeons and Dragons' designer has left – at most – the choice of four or five endings. We know that when we are eating our popcorn and watching the movie from our cinema seat, or viewing the painting from the art gallery floor, we are firmly centred in this real-world reality. Movies and paintings may represent different realities for us to watch, but they are just as much a part of our normal, consensus-built, everyday reality as are the football matches we watch, the shoes we wear or the food we eat.

But users of Virtual Reality are in two realities at one and the same time. They bring their physical bodies into a virtual world, dedicating completely their eyes, ears and hands to the surroundings and activity in Virtual Reality. But the users' feet (or bottoms) are firmly planted in 'real' reality. And should there be a great deal of cabling, the user must be aware of being in both realities, lest there be an almighty fall. So their mind, as well as their body, has to be in both worlds at the same time – for at least some of the time. American scientist and educationalist Meredith Bricken is one of the researchers who considers the essence of Virtual Reality to be inclusion – and that the virtual world exists as real. She has developed a nice analogy to explain the distinction between viewing and inclusion.

'Viewing 3-D graphics on a 2-D screen is like looking into the ocean from a glass-bottom boat. We see through the window into the environment; we experience being on the boat. Looking into a virtual world on a stereographic screen is like snorkelling. We are at the boundary of a three-dimensional environment, seeing into its depths from its edge; we experience being on the surface of the sea. Using a 3-D display with a computerised glove allows us to reach through the surface to touch objects within our grasp, while viewing our activity from outside the environment; our hands dabble in shallow water. Entering the multi-sensory world of VR is like wearing scuba gear and diving deep into the sea. By immersing ourselves in the underwater environment, moving among the reefs, listening to the whale song, picking up shells to examine, and conversing with other divers, we participate fully in the experience of exploring the ocean. We're there.'

Gabriel Olfeisch, Emeritus Professor of Educational Technology at Howard University, puts the same point another way: 'As long as you can see the screen, you are not in Virtual Reality. When the screen disappears and you see an imaginary scene – then you are in Virtual Reality.' Of course there are different views of VR. John Perry Barlow has a different slant: 'Maybe Virtual Reality is just another expression of what may be the third oldest human urge, the desire to have visions. Maybe we want to get high.'

How lifelike must a world be to be accepted as real? There are two ways of looking at this. Brenda Laurel, an actress and VR scientist, believes 'Photorealism is a false grail,' while Michael Heim, an American philosopher concerned with the metaphysical aspects of VR, reinforces this view. He believes that a virtual world *needs* to be not quite real, or it will lessen its pull on the imagination. He has written, 'A virtual world can be virtual only as long as we contrast it with the real (anchored) world.' Other commentators have argued there is no point in reproducing the real world precisely and faithfully because the real one will always be preferable to a copy. They go on to suggest that Virtual Reality *should* be unlifelike. However, this may be a rationalisation of the current state of the art. Present-day VR graphic worlds are primitive, like living in a film cartoon; *Roger Rabbit* without Bob Hoskins, as it were. But this makes them not one whit less believable to users – at the time.

The other approach to believability is from the cognitive researcher's perspective. This suggests it is possible to convey as much information with symbols as it is with accurate reproductions; and their job is to find out what these symbols are. As Jaron Lanier says, 'The secret to Virtual Reality is that the brain wants to fill in illusions.' He cites the effect of placing two widely spaced buzzers on the skin; at certain frequencies a 'phantom' third buzzer becomes apparent, somewhere between the two. The brain can be fooled, and it is the intensity of the VR experiences that is capable of stimulating the fooling. But if only parts of these virtual worlds looked more real than other parts, then there would be a bad acceptance problem. Consistency is everything.

What do we miss in Virtual Reality that we get in real reality? In theory, very little. Although systems are still in their infancy we must assume that, at the least, the technology will continue to develop along the research lines laid out today. As we have seen, this means that we should be able to read, meet, touch, feel, talk, work, and play in virtual worlds. And if we can meet and touch, then there is little doubt some of us will fall in love, at the least we will make friends – and virtual enemies. VR will act as latter-day telephone 'chat-lines' or even matchmaking bureaux. With the right psychedelic programs we should be able to get high, and we will surely be able to pray, although to what is another matter. We will miss some physical things: eating, cleaning our teeth, getting drunk or sick, and voiding. And although we will be able to take virtual showers we shall remain as dirty as ever.

We shall also miss danger. The real world is a dangerous place. Virtual Reality, no matter how lifelike, is not. Or is it? It appears impossible to be physically mugged or murdered in VR, a point endorsed by Heim. 'It [VR] can offer total safety, like the law of sanctuary in religious cultures,' he claims. But sanctuary from what? Physical mugging is one thing, but what about psychological mugging? How would abuse of our virtual persona affect the real one? And what about viruses capable of distorting your virtual world – and your virtual self? However, Lanier goes further than Heim. As a typically New Age Californian, he argues that in Virtual Reality 'Good energy becomes creatively beautiful, bad energy is channelled harmlessly.' In other words he believes that VR can only *add* to the sum total of human happiness, because virtual bad thoughts and deeds have zero real outcomes in Virtual Reality. While this has a superficial plausibility, there are serious flaws in the argument.

What can we do in Virtual Reality that we cannot do in

real reality? For a start we can be someone, or something, else. In VR we can choose to represent ourselves as anything we wish – a lobster or a book-end, a drumstick or Saturn. In effect we can live an infinite number of aliases, and, as Lanier points out, the choice of alias will say a great deal about the user. We shall also be able to visualise the unvisualisable – listening to electron densities; be in places we could never have seen – like the centre of the Earth; or see and feel sounds, good moods and bad vibes. It is truly the technology of miracles and dreams.

And in a strange sort of way, Virtual Reality has more realness than real reality; after all we start with a vacuum as it were, and have to program solid walls, transparent windows, or the effects of pressure into our virtual world. So to a very large extent VR allows us (or at least the designers of the virtual worlds) to play God. And as we have seen we can make a virtual world behave oddly. We can make water solid, and solids fluid; we can imbue inanimate objects (chairs, lamps, engines) with an intelligent life of their own. We can invent animals, singing textures, clever colours or fairies. This divine role can engender a sense of disquiet, as can the feeling that some of us may be tempted to hide in VR; after all we cannot make of our real world whatever we wish to make of it. Virtual Reality may turn out to be a great deal more comfortable than our own imperfect reality.

11

Background to the Future

Jonathan Waldern of W Industries has said that if you want
to know about the future, look to science fiction writers,
not scientists. If he is correct we are probably in for a miser-
able time. Most science fiction writers invent worlds where
humankind, if it exists, is either struggling against or trying
to cope with the consequences of a technological armaged-
don. Repression is rife, dictatorship the norm, a depressed
underclass an essential ingredient, and all the worst human
vices have submerged the better elements of human nature
– except for the hero of course.

Yet some science fiction can be looked upon as informed
prediction. From Leonardo, through Verne and Wells to
Clarke, there have been thinkers and writers whose ability
to foretell coming technologies has been impressive. It is
being argued that William Gibson should be added to the
list for his concept of Cyberspace, which some people are
interpreting as the practical extrapolation of Virtual Reality.
Should even half of Cyberspace (the ultimate global computer information network) come true, VR will have proved
to be a truly remarkable technology, a fitting addition to the
list of technologies which changed the world. The wheel,
compasses, printing, telescopes, steam power, electricity,
internal combustion engines, telephones, flight, radio, antibiotics, nuclear power and computers are just some of them.
And each trailed behind it a host of other key developments
which shaped society, from social attitudes to city growth,

language, and the development of democracy. Virtual Reality trails behind computers, but is linked to many of the others.

Everything changes, even the Rockies may crumble, but while erosion is measured in millions of years the material world as created by humans changes very quickly. Dynasties come and go, empires, democracies, cities, dictators, even federations flourish, then wither. Few last long. From Zimbabwe to Athens, and from Cortez to Lenin, fixed ideas and structures have had a finite life. Yet it is odds-on that the people living in those times believed their way of life was for ever. When Albert, the Prince Consort, first saw the electric bell at the Crystal Palace exhibition he said that civilisation and science could go no further. We must not fall into the same trap of thinking that when VR comes into everyday use it will hit the world as it is today. Indeed current trends suggest strongly that there will be considerable changes, many of which will act as forces in favour of Virtual Reality development. Let us look at one example in some detail: the car.

If you were to ask someone today whether they would consider using a virtual shopping facility, you would probably get a sad shake of the head and a response such as, 'Why bother, it's easy to do ordinary shopping?' But in ten or twenty years' time, when Virtual Reality has developed, the answer may be different. The fate of the car is crucial to the future of cities. Left to market forces, traffic congestion will worsen – Los Angeles and Cairo might well prove to be the models of tomorrow. Certainly, until now, attempts to control urban motor vehicles have foundered on the twin rocks of personal convenience and vested commercial interests. Only in dictatorships has this problem been solved, and then because cars have been made unavailable!

Paradoxically, the pollution caused by motor vehicles, which is being used so imaginatively by the anti-car lobby as the *raison d'être* for car restrictions, is ultimately likely to create a surge in their city use. Car manufacturers will face ruin if they cannot produce a viable non-polluting product, so over the next five years they will: powered by electric motors. Despite the fact that output and consequent pollution from power stations will increase, the electric car will be welcomed. With the stigma and direct pollution removed, and with what appears to be much lower initial and running costs, it is probable that more electric cars will be used than their old petrol and diesel counterparts. Jams will become worse and travel in cities take longer.

For a variety of economic and social reasons, modern shopping patterns built around car use are here to stay. Whether in French hypermarkets, American malls or British shopping centres, consumers do a weekly round, and the resulting bulk is exceptionally difficult to load and carry on public transport, especially if there are young children in tow. But if traffic gets heavier, and alternative virtual supermarkets are available, people will prefer not to brave the streets. Indeed, store owners themselves will wish to reduce the transport and manpower costs of daily stocking. Virtual shopping with cavernous local warehouses in inexpensive locations which need to be filled less frequently, and bicycle deliveries – a combination of the very new and very old – will suit both sides of the check-out.

But let us project ourselves further into the future. Cars are part and parcel of city life. The school round is now a great European tradition, and will remain so unless neighbourhood schooling is made compulsory. Even then the awareness of molestation or assault will mean that parents will prefer their children to travel in cars rather than to walk. But the consequently clogged roads delay commuters, and commuting is a pain, whether by car or public transport. But as stress awareness improves, so commuters will realise the harm they are doing to themselves. And as employers do their sums on big city payments and the cost of a strained, tired workforce, more often than not arriving late, the pressure will be on to increase telecommuting. Even today, the decentralisation of large companies, such as the Prudential and Rank Xerox, is as much about improving staff productivity as it is about the financial imperative of selling valuable city centre sites. Using Virtual Reality on Kreuger or Mandala lines, people will be able to virtually meet, talk and interact. Telecommuting will become more viable, and considerably more sociable, than just sitting at the far end of a modem.

But what will the city itself be like? Let us take the car/telecommuting argument a stage further. If the number of commuters decreases, public transport revenue will decline, as will the transport services, unless subsidies are paid. Commercial buildings will be left empty, and their prices and local taxes will fall. Also the shops, restaurants, pubs, theatres and cinemas that relied on commuters' patronage will lose custom, and may have to close. Property values will fall, multi-occupation will take over, poverty and crime will rise, education will worsen, and more city dwellers will move to the suburbs. In turn this will reduce amenities, further reducing the local tax base; transport subsidies will be out of the question. Visitors will find the city less attractive, more shops will shut so fewer workers will commute on worsening public transport. More shops and facilities will close, so even fewer people will visit the city – and so on.

VR telecommuting and teleconferencing will become even more attractive, reinforcing the decline, and the temptation to tax these virtually real services might prove irresistible to both local and central government as they attempt to make up the shortfall in revenues. Of course there is a point at which the city will be so cheap that people and trade will return, but that will depend on national and international cost comparisons. Many large cities, from New York to Tokyo and from Los Angeles to Paris, are starting to suffer in this way.

Cities in the developing countries are also struggling, but for a different reason. The birth rate is so great that the expansions of Mexico City, Cairo and Sao Paulo look endless. But will it be thought wrong to holiday in developing countries to see these, and other cities, on the grounds that we are subverting traditional cultures? There is certainly a sociological politically correct school of thought pushing hard in this direction. Indeed, concern for the physical environment will probably be so high in the future that tourism will become an antisocial activity because of the fossil fuel consumption of planes for no good reason. And from the tourist's standpoint why queue for hours to visit churches or museums, get crushed in airports, squashed in tourist class and run a real risk of getting stomach trouble, robbed, mugged, or worse? The point will be made that with Virtual Reality, you don't need to go anyway.

But it is not only holidays. Animal rights campaigns have spread around America and Europe, albeit in some countries more than others. Zoos in particular have been targeted. There would appear to be a danger that Virtual Reality will be used by campaigners to suggest that such institutions are not only barbaric, but also redundant; indeed the organisers of the proposed virtual zoo in Leicester make this very point. Will zoos be able to survive such a campaign?

But Virtual Reality will also be appearing at a time when people are becoming more isolated, and driven back on to their own resources. The mass use of television in the 1960s and 1970s converted the home into the primary leisure centre. This was subsequently reinforced by the growth in take-away and convenience foods, home drinking, increased home ownership and DIY. And, according to government social statistics, more people in Britain live alone than ever before. Furthermore, people prefer to use cars rather than buses, while supermarket shopping also limits social contact. Virtual Reality could reinforce this process – or it could virtually reverse it. Certainly more people will work, shop and play from home, but they will also be able to meet, talk and touch their colleagues in a virtual world.

Virtual Reality will impinge on an industrialised world where real reality is changing rapidly, especially in the home and the workplace. More women will be entering employment, many on a part-time or temporary basis. Both marriage and divorce will be increasing, with more children born out of wedlock. The average age will be rising as the percentage of over eighty-year-olds increases. And where will the jobs be? With each passing year more people will be coming on to the labour market, especially from developing countries and Eastern Europe. But we shall still be introducing robots, computers and other labour-saving devices. Productivity will be rising, but so will long-term unemployment in the OECD countries, severely affecting the developing nations. Who will care for the unemployed or the single parents? How will we fund the welfare of elderly people? Will there be a temptation to see VR as the purveyor of the less nutritional half of 'bread and circuses'?

It will also be a contracting world. History suggests that global recessions are difficult to shift without a major war, while economic austerity breeds an austerity of ambition. In the 1980s greed was good, profits made the world go round and investment was available – for almost anything, and in hindsight almost anyone. But now, in the 1990s, investment is low. Where is a supersonic plane to replace Concorde? Still on a drawing board, like Hotol. Plans for longer-term infrastructure changes in Africa, Asia and South and Central America have been suspended. There is no equivalent of a Chunnel, an Aswan or even a North Sea Gas conversion. Circumstances such as these determine people's mind-sets, how they feel and act. Virtual Reality could well be seen, and act, as a psychological saviour.

It goes without saying that VR will be only one of a raft of technological changes. We have touched on the electric car, but there are other energy sources, especially hydrogen. Recent research suggests the possibility of nuclear fusion within the next century. Biotechnology and genetic engineering are both likely to revolutionise health care, as well as food production and processing, perhaps even engineering. Nano-technology may well be under way in the twenty-first century. We might well be running our first space ship on hydrogen atoms, and making our cars out of new and exciting ceramics. We could be living on food substitutes, wearing clothes that change colour with our moods and living in one hundred per cent insulated, solar-powered houses, using magnetic fields to store electricity.

Political patronage could either stimulate or slow Virtual Reality development, but which will be the more likely is difficult to judge in these unstable, decentralising times. As the European Community increases its power the Scots, Basques and Calabrians demand autonomy. While English is accepted as the international business language, the Welsh, Catalans and Bretons want theirs to be taught in schools. A tenuous Commonwealth of quarrelling republics has replaced the Soviet Union, and although Yugoslavia is a failed post-war experiment, Slovenia might well find comfort with Austria, both within the EC. And will Ethiopia divide? Will Cambodia survive? VR is arriving at a time when borders, even countries, are becoming virtual themselves.

One superpower has fallen, but it is inconceivable that the world will allow America to monopolise power and influence for long. So who will take over from the USSR? Russia will be too poor. Will it be the European Community? Or Japan? Or China? Or will the world divide on the long-touted north-south lines? Or will the divisions be religious – Islam versus the rest perhaps? One thing is certain: new powers, aspirations, and nationalisms will be filling the stage on which Virtual Reality will star.

While this may not be as gloomy a future as most science fiction writers predict, it is not that far away. There will be more people with a claim on relatively static resources, and far fewer jobs available. Nations will split and nationalisms create wars. There will be technologies that few people will understand, and even fewer will be able to control. Cities will be decaying, and people will be searching for new roles in changing societies. Virtual Reality, with its intuitive base and ability to use everyday analogues, will be a godsend to millions in such circumstances.

The fate of the space industry is an apt metaphor for the end of the twentieth century. For hundreds of years visionaries visited space in their dreams. Then it happened, and for thirty years it was the frontier, to be explored and conquered. But now other priorities loom. We look inward and not outward. In the Soviet Union, time and roubles have run out, and even in America, NASA is short of funds. It seems the perfect time for Virtual Reality. Instead of exploring the outer universe, we shall substitute tours of inner universes; and Virtual Reality will generate these worlds, whether grounded in Earth itself, or within ourselves.

What has all this to do with Virtual Reality? Well, it is the first inclusive electronic communication medium that emphasises pictures and symbols rather than text. It is a perfect medium through which to communicate in what will be difficult times, especially for the less developed countries. Common symbols will emphasise common humanity, expose common difficulties and help with common solutions. VR requires less sophistication to use than other computing media. Unlike television it is interactive, it needs less basic education to understand than newspapers or journals, and it can transmit a universal 'language'. What is more, in theory, it will be transmittable to ordinary people without the filters and gloss of government and broadcasting authorities. It is the hope for the next century. It may indeed afford glimpses of heaven.

12

The VR Industry, the Mafia, True Systems, and Viruses

The San Francisco Virtual Reality conference organised by Meckler in September 1991 was a slightly downbeat affair. A great deal of mea culpa was in the air as pioneers confessed to having over-sold the VR idea in their heady early interviews. Virtual sex had taken the headlines from virtuous virtual worlds, while press handouts explaining how VR was helping disabled people had been spiked in favour of stories on how to take a snorkelling holiday on the Great Barrier Reef while sitting in a room in ice-bound Detroit. Jaron Lanier donned his virtual hair-shirt and admitted that he, among others, had raised expectations of quality too high – too soon. And it was felt by several speakers that there was no such thing as a Virtual Reality industry, it was far too early in its development.

The depressed mood merely proved to be a healthy pause for reflection after much helter-skelter expansion. By December enough products were coming on to the market to constitute a pre-fledgling industry; that is to say the egg was definitely cracking. Computers especially designed to carry the more sophisticated VR programs were on the market. New improved head-sets, optics, trackers, gloves, mice and wands were available. Games were a big attraction in arcades, and VR watchers knew they were about to get into the home. But, most important of all, users and markets existed. The specialist companies such as VPL, W Industries and Division were selling their test rigs to mainstream companies rather than to each other. Some companies, like BT in the telecommunications field, had purchased several rigs to experiment with and evaluate. At the same time software houses such as Sense8 and Dimension International were selling their virtual world kits to be used with the test rigs. And Lanier himself was something of a hero in Japan, where many of his VPL products were selling to large company labs. Both the selling and leasing of arcade games was burgeoning, home DIY toolkits were selling well, and the market for gossip, literature or indeed any information, was bounding away.

Where there are markets there are distribution systems. By early 1992 these were getting established. Alliances were forming between 'rival' companies to sell each other's products. Licences to produce, use and retail VR equipment and software were being granted. VPL, Sense8 and Mandala, for example, had agents around America and in Europe, VPL especially in Germany. Showrooms were being organised; one company in London, Virtual S, set up software, testing and design together under one roof. Consultancies such as Virtual Presence, with ten employees a substantial employer in the British VR industry, were fully occupied – despite the recession. Agents and agencies were emerging and lawyers were drawing up contracts and patents. The latest VR directory, the *Virtual Reality Sourcebook*, has over three hundred listings.

At the San Francisco conference Dr Latta of 4th Wave suggested there would have to be one hundred thousand users before VR could be taken seriously. But with the emphasis of the conference being so determinedly American, Latta did not know that Dimension International had already sold around this number of their 3-D Toolkits, which were being used in homes around Britain. But in one sense Latta was correct. Virtual Reality will only come of age when it proves itself to be both useful and commercially viable. Clever demonstrations are one thing, but reliability, robustness and relevance are what is needed.

The key to understanding where VR can go as an industry

is to remember that it is basically an enabling technology. It will be used with other technologies to make things happen, so in one sense it can be treated as a component, while in another it is a product in its own right. With these dual roles it is similar to radio, itself a consumer product and also something that is included as a part of other products and systems.

There are five sets of players in the current Virtual Reality game: the companies manufacturing VR equipment; the companies buying that equipment for their own purposes; the research institutes in universities or elsewhere which are researching, producing and inventing new products and uses (simultaneously); the peripheral players, the agencies, consultants, lawyers, journalists and accountants who are oiling the slowly turning wheels; and last, but most definitely not least, the producers of arcade games and software.

The present environment does not look as if it will sustain most of the participating manufacturing companies for very long. They are in a self-limiting market, selling to other companies for research purposes. If these particular lines of research come to nothing, interest, and with it the prospect for future sales, will wane at that particular company. If the research succeeds, there is no guarantee that the producer will get future contracts for its equipment: the company doing the research may well intend to produce its own. Today's Visual Reality manufacturing companies are, by any definition, small. None of them is large enough, or has a broad enough base to survive more than a single major set-back. More to the point, none is in a position to resist a determined predator, in large part because too many depend on the ability, even genius, of one or two high-profile people.

The exceptions are arcade games producers, like W Industries, and the research establishments such as HITLab and UNC in America, IMC in Germany and ARRC in England. The funding and backing of these labs will enable them to cast their net wider than the manufacturers. They are in two fields at one and the same time: the development of the technology itself, and the development of products based on it. This will enable them to approach different, potentially large, stable markets. Each lab has formed separate trading companies, and these are in a far stronger position than the manufacturers. Both the financial security and logistical support given by single sponsors, consortia or government involvement allow these labs the unaccustomed Western luxury of taking a longer view, and weathering the inevitable short-term blips.

The electronic arcade game and slot machine industry depends on attracting new young players with state-of-the-art technology. Even in a recession new arcade games are welcomed eagerly, just as the movies prospered greatly in the 1930s depression. The potential market for VR arcade games is very large, and that for home games exceptionally large. For this reason it is highly likely that Japanese and American producers will be in the market in some numbers. With the media concentrating its attention on games, many VR watchers and producers such as Jonathan Waldern believe that entertainment will drive VR into its next phase. To the extent that it has users, and will generate a market, sales and income before most of the other VR interests, this is correct. But whether this constitutes a driving force is quite another matter.

A driving force acts either on innovation or finance. As we have already seen, less computer power is needed for the relatively limited range of options in arcade games than, for example, that needed to manipulate molecules. Nor is accuracy necessary in arcade games; visual and aural approximations are sufficient. So on the basic computer side, games are unlikely to be an innovative force; the precision needed in sophisticated scientific and military projects will drive VR technology. However, games may help develop visual displays, helmet and hand control design, the robustness and the packaging of the entire product.

However, unless there are unexpected changes, the income generated by arcade and other games will not be ploughed back into the general VR industry. This is because companies working with Virtual Reality have divided into a games and non-games sector at an early stage, the exception being W Industries which has a 'test-bed' presence, and could plough games profits back to develop its Virtuality computer for other uses. However, it seems highly unlikely that this route will be followed in America or the rest of Europe, where companies involved in games tend to be dedicated exclusively to the entertainment sector.

But there is a darker side to the arcade games sector. In both America and Japan it is more famous for its connections with organised crime than its social and moral commitment. VR will make a lot of money for someone, somewhere, and we would be naïve to assume that the Mafia, Yakusa, Triads, or other criminal fraternities will ignore its potential.

There are many ways in which criminals will be able to muscle in. Obviously the physical protection racket will be based on the fact that while VR machines cost far more than most other arcade games, they are only as fireproof, water- or hatchet-resistant as the cheapest of them. It is also possible to threaten users. But the possibility of a really nasty virus would do just as nicely, thank you, and with far less chance of getting caught. Of course the Mafia, or Yakusa, could take the opposite tack. It only takes a single wholesaler or leasing company to put the squeeze on arcade owners for them to feel obligated to take that party's VR machines in the first instance.

Gambling has long been a favourite source of revenue for criminals, be they organised or freelance; Las Vegas-style slot machines have literally proved to be moneyspinners. However, even a casual inspection of those gambling halls reveals an ageing population; Compo and Foggy would feel at home. Few youngsters are playing. But Virtual Reality gambling machines would remedy that. Not only would they appeal to younger players, they would add almost total immersion, and there are good reasons for believing that vulnerable personalities may find Virtual Reality addictive. For such people, immersion in a VR 'slot' could be tantamount to picking their pockets. VR could deliver gangsters a considerable extra gambling dividend without having to squeeze more blood from the same set of rhinestones.

Crime battens on to human urges and weaknesses. Gambling, sex, porn, booze, and drugs have always been hugely profitable for the underworld. The moral guardians of our societies never seem to learn. Time after time they frame laws to restrict the supply of these things – so leaving a large unsatisfied demand. But as soon as the supply of any product or service is made illegal, criminals will move in as suppliers to fill the vacuum. There will be many more people waiting to supply (and to use) Virtual Reality in these ways than planning to help disabled people; indeed, given the way of the world the disabled may lag well behind. Undoubtedly, organised crime will find VR a profitable milch cow in several ways, but few, if any, of the profits they generate will find their way back to the VR industry.

At present this entire Virtual Reality mélange exists in an untidy world, where common standards, one of the key hallmarks of an industry, do not exist. Instead companies, individuals and labs are doing their own thing, inventing their own platforms, protocols and procedures. VPL is trying to codify standards around its own products by offering to trade patents to companies which agree to work with VPL standards. Imaginative it may be, but such is the level of distrust in the industry that, unfortunately, this offer is as likely to be spurned as accepted.

Market research is not being taken seriously either, with the exception of the entertainment industry. Certainly COMPEC, a consortium to study the market for VR entertainments, stands out. Not only does it buck the amateur trend, it is also based on solid research foundations and analysis. But for the most part ideas are being followed because they sound like good ones to the manufacturer or developer, not because a need has been demonstrated. And should the need be there, the potential demand (a very different matter which has more to do with price, style, fashion, delivery and ease of use) is being ignored. Far too many products are being led by the technology rather than by consumers or users.

It is never wise to predict what will happen to companies.

More money has been lost 'investing' in stock market tips than has ever passed into bookmakers' hands. Yet the history of industries based on new technologies does not augur well for small pioneers. The motor car is a classic example. Small enterprises producing specialist cars proliferated initially. Although rationalisation started early with Ford, it developed in waves post 1945, until the giant manufacturers absorbed most of the competition. But since those times even larger, more powerful multinational companies have come to dominate, so that while some of the small VR manufacturers may survive intact, the odds are stacked against them.

Given the precedents it is inevitable that the bigger companies are biding their time to see how things turn out before moving in. But in Japan they are doing more than just waiting. With government support, both financial and moral, companies, universities and research institutes are sponsoring different lines of development, especially in the telecommunications area. It would appear that most of them are intending to improve on, or extend, lines of action already being taken in America or Europe. Although the possibility remains that original research is being undertaken behind closed doors, most of the effort seems commercially oriented, rather than a search after new scientific truths.

Japanese Confucian/Shinto culture encourages the very long-term view. Most Western corporation managing directors are proud of themselves if they can unveil a five-year strategic plan. Recently Matsushita executives spoke at a Harvard seminar outlining their five-hundred-year plan. This attitude means the Japanese are prepared to put money into Virtual Reality development without necessarily needing quick returns. Companies as large as NTT and Sony are happy to bring products to the market only when they are ready. If in the process they bankrupt existing smaller companies, wherever they may be, that is business.

As we have already seen British and European computer manufacturers are also studying the potential of Virtual Reality, but have neither the government support nor the sense of urgency needed to bring out products before their Japanese competitors. In time-honoured Western style they are more likely to acquire expertise, products and markets through takeovers. But software producers are in a different position. True, they will have copyright and copying problems and they may well fall victim to viruses, but they are likely to remain in small units, unless a super-game takes off. The same argument applies to graphics, animation and consultancy houses. Again, in this area, the Japanese are the more likely to take the bull by the horns and devote resources to games machines and software. People who study the Japanese home market have pointed out that there is a double incentive: Japanese people are fascinated by gadgets; they live in confined spaces, in a country where open spaces for recreation are scarce and houses and flats are compact. Virtual Reality equipment, being small and deserving of the soubriquet 'gadget', fits their needs perfectly.

The Virtual Reality industry is unlikely to settle down before the start of the next century. And when it does, the electronics, computing and telecommunications megacorporations will be in there, battling it out. This is because so many of the most profitable applications need a great deal of infrastructure investment and development. Virtuphones and virtual teleconferencing need land lines and air-space, capital equipment and a large number of subscribers to be commercially worth while. Small companies, however innovative and well run, cannot raise such large amounts of operating capital. A major reason for delays in this shakeout will be the combination of Virtual Reality with other technologies, bringing in new companies, new ideas and new motives.

If Virtual Reality can be accused of being a misleading name, then the inventor of the title Artificial Intelligence (AI) should be prosecuted for false pretences. It implies a substitute human intelligence, while it is really no such thing. Yes, the systems can learn, and yes, they can take decisions on the basis of what they have learned, but the subtlety of human intelligence is just not there. Smart, knowledge-based or expert systems would be more accurate names. VR and AI spell *vrai*, 'true', in French, and it is true that when coupled they will make a formidable team, especially in the information, library and teaching worlds, and indeed as the working basis for virtual adventure and fantasy games. We shall name them VRAI systems.

Imagine, some time in the future, a virtual library run with a VRAI system. Blocs of books, journals and manuscripts glow as you approach. If you have asked for references on, say, recent Italian research on the radioactive compounds of calcium, a combination of pulsing different coloured lights and encouraging sounds will guide you from inorganic chemistry, to radioactivity, on to calcium compounds, and then to Italian research. The papers you want are highlighted and you can have them printed, or presented in analogue form. The system will remember the question and the route it took to answer it, and store it for future use. Indeed, your route to the Italian research will probably have been refined by the VRAI system using its memory of a previous enquiry, which may have had little to do with radioactivity, calcium or Italy. If you ask immediately for French research on calcium the system will link the two routes together for future reference, or will amend existing routes.

Libraries and information sourcing, retrieval and presentation will be some of the things VRAI systems do best. They will shortcut existing methods and, if required, present the information in ways that appeal to the senses, rather than in indigestible wodges of text and data. Forests of information would be an apt visual metaphor, considering the trees being saved by this method. In recent years there has been a perceptible shift from data to information need. As the interrelationships within the real world are perceived to be more complex, so companies, governments and scholars alike have come to realise the value of apparently random information in their decision-making processes. But this carries the rarely mentioned, but all too real, danger of being *overinformed*. This condition can cause paralysis systems (paralysis by analysis) comparable to those caused by the better-known underinformed syndrome. Systems which can get information abstracts in the shortest time possible are worth their weight in silver; those that can find their way around networks, sniffing out nuggets of 'lateral information' will be worth their weight in gold.

Artificial intelligence has been hibernating. Lauded to the skies in the 1980s it fell to earth because it was not enough like human intelligence. Nevertheless, smart systems are used widely in diagnostic medicine, military decision-making and stock market dealer programs. Professor Donald Michie of the Turing Institute had no doubt about their potency when he said, 'If a machine gets very complicated, it becomes pointless to argue whether it has a mind of its own. It so obviously does that you had better get on good terms with it and shut up about the metaphysics.' But they are not infallible; externalities can throw them completely off course, and they helped create 'Black Monday'. VRAI systems will allow us to glimpse heaven as well as provide visions of hell. Heaven will be approached through virtual libraries and celestial intensive care units. Heaven will be clever futures universes made of virtual oceans, prairies and cloud formations, telling you what stocks to trade, and when (with built-in fail-safe triggers). Heaven will also be strategies to solve Third World famine, with images of deserts and the sounds of laments. But hell is to be found in a different place altogether.

A crucial feature of AI systems is that they retain the footprints – perhaps brainwaves would be a better description – of those who have used them before. Suppose a psychotic has been using a VRAI system, which responded to that person's aberrant thought patterns. These will now be built into the system's memory and behaviour patterns. If a 'normal' person stumbled on to a path which triggers these psychotic routes, the 'normal' user may become trapped in an irrational, even dangerous web. Remember, this is neither an everyday computer nor a television-style visual experience. It is highly intense, highly charged and almost totally inclusive. It cannot be shrugged off like a horror movie, reading a sci-fi book, or scanning a spreadsheet. You are 'in' that psychotic world. What makes matters so much worse is that the 'normal' user will have no way of knowing who has used the system previously, so they will believe the system is normal. If this is a maze, or a fantasy game, then young, impressionable, perhaps vulnerable people could be playing. If it is a self-generating graphics program, some of the animation or pictures may be disturbing in the extreme. And if it is a networked information program, countless numbers of people could be involved. It is a form of human-carried computer virus.

If information is to be a key VRAI function, then communications will be the bread and butter of Virtual Reality. Networks and networking are the business techniques of the future, but being played with today. They are the framework over which the concept of Cyberspace will be draped, before being fleshed out with databases and communication terminals. Fibre-optic cabling, of one sort or another, will be the commonest form of world telecom cabling by the mid twenty-first century. This worldwide Integrated Broadband Communication Network (IBCN) has started: 121 undersea fibre-optic cables are running, or are planned. The reason for the installation is the very high running costs of conventional cable – it is more profitable for the carriers to instal fibre over the longer term, even taking into account the initial capital costs, which are falling. These networks will be carrying videophone, computer and virtuphone traffic, as well as teleconferencing. It has to be remembered that almost the entire East European network needs to be replaced, while much of Africa, Asia and South America will be starting from scratch, so the additional cost of fibre, when related to total costs, is relatively low.

Before we reach this point, however, debates will have to be won, and major decisions taken. At present the proposal is to make use of data 'compression' techniques to get High Definition Television (HDTV) into homes, rather than waiting for fibre optic cabling. While the motives are laudable ('compression' will also allow for some degree of inter-activity), it may well stunt the growth of fibre, and VR, in the medium term. Yet the difference between fibre and compression is staggering. In the lab, a single fibre can send thirty-two *billion* bits of information (gigabits) per second, equivalent to two thousand books of four hundred pages each or four hundred thousand simultaneous phone calls over a distance of thirty-five miles. In the outside world ten gigabits is around the corner. But even with 'compression' copper cabling can deliver only twenty megabits. The band-width case for fibre is overwhelming.

Information is the oil of the new world order. It fuels business and government, academia and research. Security, health and welfare all depend on it. Within the foreseeable future countries will be as reliant upon information net-works for their continued well-being as companies are today on their computers, or the global transport service is on gasoline. But with such dependence come dangers – and there are two. Networks are vulnerable to accidents, terror-ists or criminals. No cable is bomb-proof. No computer-driven system proof against either complete power failure or complete infestation by viruses.

Back-up systems are prudent, but increasingly they will have to be dispersed over several countries. In any event a halfway competent terrorist (or criminal) organisation should be able to sabotage the network rather than systems themselves. In this way they can kill several birds with one stone. Viruses are another matter. It is suggested that some originate in Bulgaria, but that is of little significance – if they came from Belgravia they would be just as lethal. The dam-age they can do is incalculable. But in VR there is an added dimension, which we shall approach when we consider the effects on individuals. Viruses can be made to affect a virtual world. They may not only act on the shapes and contours of users, but also the user's behaviour in relation to the world, indeed the behaviour of other objects in it. The first virtual 'deaths' will come when viruses destroy virtual images of people in precisely the same way as they destroy other digital representations, for example conventional computer data and text.

The second danger is inequality. Some citizens, even some entire countries, may find themselves with insufficient information to compete, or in extremes even survive, because the information network is not available to them. With its visual and aural analogues, images, and sense of touch, Virtual Reality information is capable of communicating on a universal basis. Of course the graphics and symbolism would both have to be carefully created to be broadly cross-cultural, but with the addition of culturally specific images when required. As this property makes Virtual Reality capable of being used by people who have but a limited education, being isolated from the network will be particularly damaging, and illiteracy is increasing across Asia and South America. Virtual Reality's ability to demonstrate how to find and dig a well and make a simple pump, how to use various pieces of equipment, or how to promote hygiene and disease control will be irreplaceable, as will its ability to superimpose the image of a perfect product on work in progress. Reading skills, or a knowledge of English, German or Japanese, will not be needed. It will be hands-on training without having to export much in the way of trainers, machinery or infrastructure. However, there will be a need for Virtual Reality engines and peripherals, as well as service engineers. UNESCO, and other international and United Nations agencies, should be evaluating its benefits and costs now. By the next century VR could be the carrier of the most widely dispersed education and training scheme ever devised, although without a network of sufficient bandwidth its delivery will be patchy.

If it does get to people who need it, existing inequalities should close: but without this education blitz, inequalities undoubtedly will widen. This is because late twentieth-century technological advances have actually reinforced global inequalities. The industrialised countries are leaving the non-industrialised further behind with each successive wave of developments. This need not be the case, it is a question of what we allow technologies to be used for. However, it is important because other new techniques will be used with VR. For example, if the University of North Carolina can manipulate pharmaceutical molecules, there is no theoretical reason why they cannot do the same with enzymes to facilitate genetic engineering. It will dramatically cut costs in this area. And why stop at this level?

Nano-technology is engineering at the atomic level. In theory it is possible, in practice it has been impossible. But Virtual Reality will allow the visualisation and manipulation of individual atoms. Virtual telepresence engineering could then prove to be the key to products as unusual as woven diamond fabric and cold liquid glass, totally biodegradable plastic and solid mercury. Nano-technology could change the world. The benefits are enormous, but it is likely they will accrue to the industrialised world almost exclusively. As things stand at present only a trickle would reach developing countries. But even nano-technological products start to pale into insignificance when nano-technology, neural networks, DNA-based biocomputers, and Virtual Reality get together. We could live in smart buildings, eat (or perhaps compute on) clever carrots, and live with artificial kidneys grafted on to self-repairing ersatz skin. We would be into a world of bionics which would transcend the human body.

Both in concept and in practice, Virtual Reality supplements present common human experience of a single physical commonly accepted reality. It is another 'big step for mankind'. But while much of the VR industry is finding its feet, and starting to test opinion in an often wary world, the entertainment side is set to roar ahead. But the litmus test of VR technology will not be taken on the Covent Garden Plaza or in a Chicago 'battletank'. It will be taken in hospital wards and school classrooms, offices and shops, aeroplanes and homes. To get anywhere near fulfilling its undoubted potential it will have to enter the very fabric of our lives, much as the telephone and television have done. And it must affect the dual-time consumers of our waking lives, work and leisure.

13

Workplaces and Playplaces

Employment is important; the work (or paid employment) ethic is very strong, so much so that people identify with their job. Ask anyone standing near you at a party what they do, and they will reply with a job title. It is so important that leisure has been relegated to the residual time left over after employment. But you have to earn your enjoyment, especially if it costs money; unemployed people who receive state benefits and who are seen paying for their fun are often objects of envy, sometimes hatred. But play is at least as important to a healthy life as is work. Even the most competitive of employers recognise that people need some time off to recharge their batteries. So anything with an impact on jobs has to be taken seriously, and anything with an impact on both jobs and leisure, like Virtual Reality, will be treated with more than a soupçon of suspicion, by employees and hedonists alike.

In the eighteenth and nineteenth centuries there was a great demand for ostlers, gas-lamp lighters and quill pen sharpeners. Workers were proud of their skills. There was a definite pecking order, with skilled men the industrial aristocrats, and when their skills became redundant, they felt it keenly. The Luddites rebelled because their weaving skills were no longer needed, not because they would be jobless. But jobs come and go, which is why we no longer need quill pen sharpeners or lamplighters; nor indeed the far more recent comptometer operators and punch card clerks. Yet, up until the 1980s parents were still able to urge their children to learn a trade so as to get security. 'It'll stand you in good stead for the rest of your life.' But that disappeared from then on when computer-driven techniques drove many skilled trades into the same oblivion as the ostler.

It is now accepted widely that jobs, and skills once developed, can no longer be followed for an entire working lifetime. And although not always practised, the need for regular retraining is now established. This is just as well, because Virtual Reality will have an impact on employment, especially in relation to skills. The bull point of VR as a computer interface is that it is easy to use. Children can be trained to use a head-mounted, gloved system in less than fifteen minutes. Its strength comes from doing what comes naturally. Anyone with normal hand-eye co-ordination can be trained to use a VR computer system. But people operating computers today would claim that they have special skills, and that other people would need considerable training to do their jobs. This will not be the case with VR interfaces. Virtual Reality will lead to de-skilling.

Does this matter? After all it will only affect a relatively small, and diminishing, number of people. But such is the strength of status, most of them will feel personally diminished – and they will not be alone. It will be no new phenomenon. Computer use has revolutionised workplaces over the past twenty years. Whole swathes of manufacturing industry, manual, skilled and semi-skilled jobs have gone, replaced by computer-driven systems, as have millions of filing clerks, printers, and middle managers. And even the second-generation computer workers, the paper tape librarians, have seen their jobs disappear. But at the same time jobs were created in computing itself, in services, in design, and in areas such as credit card companies which could not exist without data-processing.

But computers had another effect: they changed the way we did things at work. Some employees, like the typist, gained skills from the word-processor and desktop computers, but professional engineers found themselves constrained and felt undermined by the new systems. Computers also compartmentalised businesses, and with this came the fragmentation of jobs. While computers always had the capability of widening people's job horizons, in the main they closed them. And it is a sad fact that although VR is also capable of widening horizons, most jobs do not (and probably will not) allow employees the latitude to take advantage of it. Instead, today's jobs concentrate on very narrow tasks and in the work environment VR may well be used to narrow these tasks even further.

Until Virtual Reality has developed further it is difficult to pinpoint which jobs will be affected. For the most part today's computers are used only as a tool to help perform specified tasks, not for a whole job. There are exceptions: most 'typists' depend wholly on word-processors. But those jobs which involve accessing and using information, such as researchers, bank tellers, insurance clerks, air-line and other counter clerks, telephone operators, as well as some managerial and secretarial positions, will feel the changes. Not only will information be physically easier to get at, it will be presented in a more accessible form. Certainly jobs will be created for people to put the information into this form, but they will be more creative and imaginative than technical. Skills will be changing from one group of workers to another and, moreover, the skills needed will be different. This raises two further points.

One way to upgrade the work of less skilled people is by extra training. Another is to allow them to do the job by making systems more comprehensible, and this is precisely what VR will do. Relatively unskilled workers will be capable of both retrieving and interpreting information. De-skilling yet again. But this may also lead to a loss of jobs. If air-line reservations can be made using graphic representations of plane interiors, why not create a help-yourself system? Or indeed a theatre, holiday or almost any other booking system. And with a fibre network, why not do it from home, the office, the local post office or supermarket? Certainly people will be needed to produce and update the database, but with VRAI systems there will be far fewer of them than of the employees who used to interface with the public.

But if skilled workers lose, unskilled workers gain. Over the years, far too many intelligent people have done themselves scant justice in the education system, and as a result have found themselves condemned to dead-end jobs, well below their potential. Virtual Reality could be the passport to liberation for these people. It will reveal to the relatively untutored the secrets of statistics and finance, medicine and science, by stimulating their visual, aural and probably tactile imaginations. And there is no reason why they should not do at least as good a job as their better qualified predecessors. Why would employers want to employ less skilled people? Simple: in industrialised countries it keeps the wage bill down; in developing countries it makes better use of the existing labour force.

Virtual Reality will create other workforce shifts. Virtushopping, using VRAI stocking systems, will cost the jobs of the majority of supermarket shelf stackers and check-out clerks, but a legion of delivery cyclists will be created, reducing the average age of the workforce. Telepresence robots may replace many unskilled safety and security personnel with more skilled 'puppeteers'. The present-day transfer of clerical, administrative and analytical work to places such as Singapore and Malaysia where wage costs are low, troublesome unions banned and there are no limits on overtime, will increase. Working from home will allow nursing mothers, and disabled people, to compete on the labour market on equal terms.

From the moment remote terminals were available, some twenty years ago, it was clear telecommuting would use resources efficiently. To some extent it has succeeded. A survey of American workers in early 1992 showed that one and a half million telecommuted regularly and another three and a half million did some telecommuting in the course of a working month. Nevertheless human considerations have a habit of interposing themselves. The lack of regular contact with other staff had impeded satisfactory results. But three new factors are coming into play. These are the move towards smaller 'core' full-time workforces, topped up by 'peripheral' part-time and temporary employees, as and when needed; the parlous state of public transport and road systems; and the move to distributed computing which enables (or forces) managers to work from home, cars, trains or wherever.

Now, add on to these trends a Virtual Reality-based tele-conferencing system, and combine it with both video and virtuphones. It is the breakthrough. People will be able to work from remote offices, or from home, without most of the original disadvantages. And when it arrives retinal imaging will add to the illusion of inclusion. Using such a system, by the mid twenty-first century a considerable number of employees will not be attending their offices on a daily basis.

It is hard to overestimate the effect VR will have. The drawbacks that hindered telecommuting will be washed away in one deluge of technology. People will be able to meet, talk, touch, argue, demonstrate, consult, advise, sell, check body language, and in general interact with other people without having to move from their bedroom. The office, indeed the business world, will be coming to the workers rather than the workers moving to the business and office. And it will feel real.

Of course many jobs need a hands-on approach – fire-fighters, butchers and ambulance-drivers are examples – but others can be done, in part at least, from home. Typists, secretaries and administrators, computer programmers and accountants, salesforces, executives, designers, professional engineers, publishers and writers are obvious examples. But psychologists, lecturers and teachers, accountants, doctors, even lawyers and judges will be able to perform some of their duties from home, or from wherever they choose. And even if companies do not go as far as Rank Xerox, and hive off a number of their managers to become separate companies, most managers will be able to work a more flexible pattern: in an office when they need to be, through the network at other times. On the other hand it could be argued that managers will also be at the beck and call of their employers on a twenty-four-hour day, fifty-two-week year basis. In theory they can shut down their equipment, but they will be competing with some of their colleagues who will always keep it open. VR will be of undiluted benefit only to employers.

When telecommuting (virtucommuting is better) takes off, small will not only be beautiful – it will be necessary. Compact, highly cabled electronic offices will replace the large corporate office blocks which will stand only as monuments to previous business philosophies. Indeed, the days of one worker one office are numbered, unless that office is in the home. And with VR-organised distance learning and medical diagnostic techniques coming to the fore, even the number and size of college and medical buildings will decrease. Not a happy thought for the construction industry.

And if Futura, a computer-generated virtual city, complete with virtual offices and conference rooms is heaven to Professor Francesco Gardin of Artificial Reality Systems, Milan, it is sheer hell for the building trade. Professor Gardin's city exists exclusively to enable people to virtucommute. Rent a virtual Futuran room, on with the helmet and DataGlove (or use a space-ball) and have your business meetings in virtual space. The participants may be at home, on holiday, in the bath or watching television. Professor Gardin's vision is that the informality of VR, as opposed to formal teleconferencing, will allow for normal conventions to be observed. People will be able to have chats in corners and allow their body language to express their true, as opposed to verbal, intentions. The system is to be piloted in 1993, with a thousand people in the system's network.

A stroll around the remains of Pompeii or Thebes is a salutary experience. It reminds us that cities as well as people have finite spans. As trade, religions and technologies have changed so cities have risen to prominence before slipping back into obscurity. Today's picturesque French bastides and genteel English market and county towns represent a past power structure, while cities as diverse as Liverpool, Sheffield, Lille, and Detroit have been stranded by the ebbing tide of steel and car production, and now are casting around for new roles. Virtual Reality will be a catalyst for change every bit as strong as these changes were in their day.

The metamorphosis of telecommuting into virtucommuting will have deep demographic effects. Where once suburban roads, separating ranks of dormitory houses, served only as pathways for au pairs, mothers and prams, they will be transformed into the arteries of new working and living environments. Where up-market shops, restaurants and entertainments were city-centre-based, now they will serve the suburban virtucommuters. The rise of suburbia will be the mirror image of inner city decline. The cycle of urban change will have turned another degree.

Virtucommuting will have several advantages. To a point it will allow employees to pace their work, while fitting it in around their lifestyle. The point comes when virtuconferencing needs clash with personal needs, and the conference takes precedence. There will be fewer commuter journeys, expenses will be lower, time will be saved, and in many cases job satisfaction will be increased. From an employer's point of view it will cut overhead costs (office blocks can be sold), and increase efficiency. It will reduce the distractions when office relationships go sour, defuse the interminable round of office politics, and diminish the thorny problem of sexual harassment.

But how important is actual physical presence? Workplaces are social places. Friends are made at work, and without a workplace many people will be lonely; and as divorce rates soar and single-parent families predominate, so the numbers of potentially lonely people increase. Can virtual world contact make up for the loss of physical social contact? It is just too early to know whether virtual worlds will be able to make up for conventional social deprivation.

Distance learning has been more successful than telecommuting, possibly because studying can be a solitary experience. The American Sunrise, and more especially the Open Universities, have proved to be highly popular. But Virtual Reality will take education and training into another dimension. It will be able to translate hands-on techniques into virtual worlds in students' homes. It will link teacher and student, or students, in seminars, lectures, interactive demonstrations or master-classes. Whether the VR learning programs are for schools, colleges, universities, apprenticeships, refresher courses, for the physically disabled or those with learning difficulties, or indeed are just for fun and interest, imagination will be allowed full rein. Galaxies can be invented and explored, chemicals created and reactions monitored. Gaining a student's interest will be limited only by the creativity of teachers, students and virtual world designers.

Virtual Reality home learning will have many advantages. Capital equipment and material costs will be kept to a minimum. It costs comparatively little to build a virtual nuclear reactor or operating theatre, or indeed travel to Patagonia. If the wings fall off the plane you have designed, no harm is done. Indeed, as Wilde points out, 'Experience is the name people give to their mistakes,' and VR is a marvellous medium in which to make them. Students will be able to interact with experts and get into situations which otherwise would not be possible. This is particularly valuable for disabled people or others who, for various reasons, are not allowed to mix freely: women in Islamic countries for example.

There is no reason why home VR should not be the basis for a lot of regular school and college work. The attractiveness of the system itself will provide motivation for students: indeed given the educational problems on both sides of the Atlantic, VR might beat the regular classroom in this respect. While it is easy to see how history or geography can be made considerably more interesting, practical subjects such as music or design technology, art and engineering will be just as well suited. In practice as well as in theory, the direct access that VR can provide to world-famous specialists in these fields may make it *the* optimal teaching medium. And examinations can be taken at home in a virtual world. Vivas are a natural, as are many practical subjects such as chemistry or biology. The results could even be through on the same day.

But the physical aspects of how Virtual Reality affects the user have yet to be addressed. Currently, there are few bodies examining these important areas. Doubts about the psychological and social effects of VR are mirrored by physical concerns. Sitting in front of a Visual Display Unit (VDU) for lengthy periods has attracted awards from Industrial Tribunals when it has been shown to lead to damaged eyesight and postural ailments. VDUs are also suspected by some of giving unspecified radiation damage. What then will happen to people wearing helmets? Will the weight damage neck muscles? Will close proximity to LCD screens damage eyes? We do know of two problems. The first is a vertigo-type attack caused by visual perceptions not corresponding to real world sensations (a long lag time for example). This sets up a feedback loop affecting the middle ear, creating dizziness, or feelings akin to travel-sickness.

The 're-entry' problem is an inability to cope with the real world after a spell in a virtual one. It may only be a semi-joke when it is suggested that you can always tell people who have just emerged from Virtual Reality; they are the ones who try to walk through walls and cross the room by pointing with their finger. There are also allegations that at one British research establishment the number of car crashes has rocketed since work began on testing virtual cockpits, but this is impossible to verify. Certainly workers should be informed of possible dangers by employers, the European Commission (which already issues regulations on conventional VDU use) and the Health and Safety Executive (HSE). And all three bodies should sponsor research immediately into the ergonomics of VR. Pre-empting tragedies is always better than learning from them.

But if we assume that helmets and LCD screens will be superseded by more developed hardware, we merely change one set of possible hazards for another. Will retinal imaging damage the eyes? Or generating phosphenes the brain? Will prolonged exposure to adjacent electromagnetic fields affect the human body adversely? How long can we stay in a virtual world before there are side-effects, if indeed there are any? And there are other, wilder ideas. Brain implants, to communicate directly with virtual worlds, for example. Even if they proved to be more than fantasy, would you volunteer for one without copper-bottomed guarantees? Research has suggested links between externally applied electrical currents and brain tumours, so the guarantees are more likely to be lead-lined.

The European Commission is now taking another important tack. Directive 90/270/EEC issues regulations covering not only work station screens, keyboards, desks, chairs, space, noise, and heat, but also the software. In future this must be suitable and easy to use, must take into account software ergonomics, display information of suitable format and pace, and be *appropriate to the user's level of knowledge* (our italics). While the hardware regulations will need investigation, clearly these software regulations will make Virtual Reality use far more attractive – it will meet the EC's requirements on almost all counts.

But worktime for some people is providing playtime for others. From the DJs who take us through the night, to actors and stage-hands, camera-operators, night-club bouncers, dancers and theme park guides, millions of people work hard at bringing us pleasure. But for how long will they be needed? Will a combination of polluted cities with severe transport problems and the availability of Virtual Reality change these patterns out of all recognition?

If we all spent our waking hours at work, or at worthy pursuits, with no thought to entertainment or art, humankind would be the poorer. While we know that all work and no play makes Jack and Jill dull kids, it is less well known that over a third of average income is spent on leisure activities. If we had neither entertainment nor arts, not only would we be spiritually poorer, it would be an economic disaster. Leisure income is needed to keep the economy going, to create jobs, and keep the wheels of industry turning. It is only too true that dancing, holidays and generally enjoying yourself are in the national interest! And of one thing we can be certain, the entertainment industry will do its best to ensure that we all spend our fair share, and more than a small percentage of this will go on VR – even on virtual keep-fit. Randal Walser of Autodesk has suggested a virtual gym, where bicycles, treadmills and rowing machines can be used in virtual environments.

American, Japanese and now British demonstrations of three-dimensional television, using one of many techniques – most involving the wearing of glasses – are almost commonplace. And despite the drawback of flicker, it is now a question of which will be the most commercial method of replacing flat screens. The change will prove to be as radical as the introduction of colour TV, and two-dimensional viewing will be phased out as quickly as was monochrome television; the old presentation will look dull and lack credibility. But importantly 3-D TV will nudge Virtual Reality communications and entertainment that much further into the mainstream.

VR networks will carry different types of virtual world into our homes: packaged programs will allow for massive personal interaction, but where the basic scenarios have been pre-designed. These will reflect anything from drama to deep educational experiences, through documentaries, travel, pornography, game shows, quizzes and comedy. Advertising, promotional and marketing material will come directly to the home or be sent on disc. As these two varieties will be the potential moneyspinners, it can be anticipated with some confidence that we shall be subjected to a massive yet subtle marketing exercise.

Shared virtual worlds, the leisure equivalent of virtucommuting, with group interactions on Minitel lines, will act as play spaces, and be improvisation-based; as will the personal uses of virtual space for meditation, stimulation, enjoyment or sanctuary. Whether unwinding at the end of a hard day or escaping from the attentions of demanding friends or families, slipping into your own virtual world will beat alcohol or drugs. But most people will need the help of DIY virtual-world kits for their own constructions. This will lead to a wide range of kits, catering for all sorts of tastes and proclivities. But whatever the subject or the objective of the world, style will be dictated by the 'viewer' – who will be transformed into a doer. VR transmissions will be active, rather than passive. Couch potatoes have almost had their day.

Psychologists have barely started to consider the consequences of Virtual Reality, although when they have, some initial reactions have been less than favourable. They argue that human-made worlds must be more limiting than the rich experience of the real world. Therefore virtual worlds are inferior. As far as learning is concerned this must be true. Trying to learn about people, their responses and their foibles from a virtual world would be like humans trying to teach lion cubs how to hunt. But that view is taken from the perspective of someone who lives a reasonably full life. For a person with few friends or family, no job and little money, the real world may be far more limiting than a virtual world.

Yet the psychologists' initial reactions are understandable. They boil down to a very human desire to believe that a life of interacting with other people is the only one worth living; anything else is unhealthy. But it also contains more than a tatter of hair-shirt, no pain no gain philosophy; it implies that we have to take the rough with the smooth, sorrow with fun and death with life. It follows that a person with no friends, job or money will be better off struggling miserably than being immersed in a more friendly virtual world. But never before have we been able to interact with people in Virtual Reality. We just do not know whether relationships of this type are emotionally satisfying. We do not know whether virtual travel is as good, or even 'better' than real travel. All our instincts tell us that it cannot be – but these are based on pre-Virtual Reality times.

One further question must be raised. Who will design the programs? While this is important for entertainment and marketing, it is doubly so for education. The Jesuits believe a man's character can be forged by the time he is seven years old. To allow educational VR programs to be made without let or hindrance will be to abdicate responsibility for our children. This is not just about courses and their content: if lessons are as interactive and intensive as VR can make them, there is a real danger of brainwashing, and some people can be conditioned to believe anything.

In classical Greek drama, a character existed only when mentioned. The performer could be standing centre-stage, yet not be there as far as the audience was concerned. But one word from another actor was enough to recall that character from the void. The audience understood the convention. For them reality was altered the moment they entered the theatre. Disbelief was suspended in a way that seems strange in the twentieth century. Television, movies, newspaper photographs, all ground us firmly in a consensus reality. Yet even now theatre demands we suspend most of our everyday critical faculties. In drama actors speak to each other while facing the audience, while in pantomime we accept as normal women playing the male heroes, and men the female comic leads. We are used to realities other than those of the daily grind. That is why Virtual Reality will be accepted so readily as an entertainment or artistic medium, and will come to be thought the norm.

But will taking part in a home virtual theatre group replace going to conventional theatres, or indeed going out to interactive theatre performances? Will being able to tour the British Museum or Smithsonian on your VR disc deter you from travelling to the museums proper? Will a virtual Louvre make an adequate replacement for the real thing? Virtusafaris are certainly safer than real ones, virtutravel to the Acropolis is cooler, and a virtual Disneyland could save six hours' queuing – per day! There is a grim humour in the probability that new VR uses, such as interactive theatre and virtual theme park rides will have a very short real life, before being overtaken by home Virtual Reality substitutes. But will we be satisfied with these substitutes? Will we be happy in the home, relating more to virtual than to real worlds?

It is only too easy to believe that the world has always existed in its present form. But there was a time before television, indeed before radio and telephones. Whatever did we do before them? Some people say we had conversations, read books or played party games. Certainly the middle-class Victorians were great ones for family musical evenings, soirées and charades. Virtual Reality contains the possibility of a return to those days. A shared world could be a playplace; an area in which people join together to create their own amusements: plays, musical evenings, dances or whatever. The difference will be that one part of the family might be in Alice Springs, another in Buffalo, and the rest in Camden Town.

Entertainment, however, is in the eye of the beholder. Not everyone will want to reawaken family memories or be jolly. Restaurants will always be in fashion; after all virtual eating will not be too enjoyable (even if it is good for slimming) and conventional theatre and cinema may well survive. With the probable marketing blitz, it is highly likely that pre-packed programs will be the most popular adaptation of Virtual Reality. Think of a soap where viewers are also participants and, to some extent, are able to interact with the characters. Think of travel and nature programmes where you choose the route, and the viewing points, or a mystery where you play through several endings. Virtual Reality is where dreams really come true, where you score the winning cup final goal, do amazing skateboard tricks, sing at the Met, paint like Van Gogh or star in your own movie. And why not map well-known performers and keep them for posterity? In twenty years' time a movie producer will be able to 'use' the performer just as they look today, and have them playing alongside performers who were not even born when the map was made. VR will be perfect nostalgia fodder, ideal for politicians and religious leaders, and make for super interactive pornography material – yet another area of interest to organised crime.

There is also the possibility of visualising your own worlds, in Lewis Carroll fashion. VEOS software developed at the HITLab allows each user to see other users in different guises. For example user A is on a network with users B, C, and D. User A sees the others as dogs in a park; at the same time user B sees the other three as lions in the bush while user C sees A, B and D as different atoms. In other words a user can develop a personal world in which to represent other users. VR has become a looking-glass in which different users see different images of the same scene.

Looked at from the perspective of life as it is today it is difficult to believe that people will choose to give up touring the real world. But we are concerned with the future, say fifty years' time, when travel is likely to be hazardous, expensive and considered to be downright antisocial. In such circumstance VR might be the option of choice. If you can see the *Mona Lisa* from all angles or touch a *David* everywhere, with a choice of lighting arrangements, accompanied by an expert, friendly commentary, why take second best and see them imperfectly through a horde of fellow tourists in Paris and Florence? As far as VR itself is concerned, interactive art galleries will be interacted with from home, as will interactive theatre groups. By the middle of the twenty-first century, new dwellings will be designed with special movable partitions to create Virtual Reality capsules so different members of the family can be left to their own worlds. By then the home may well have become a combination of living space, workplace and playspace.

14

A New World for Women

Two well-rehearsed, and widely accepted assertions – made by both women and men – are that women will never be as good at maths as men, and they are temperamentally unsuited to computers. These represent only the tip of a generalisation iceberg of gender differences. Men have constructed their own set of wish-fulfilment attributes which describe the ideal woman, generally to do with child-bearing and home-making. On the other hand women do not object to being described as humorous, magical, intuitive, and emotional – indeed many take pride in the descriptions – or to men being described as rational, serious, pragmatic and scientific. But while they may contain an element of truth, looked at closely these generalisations describe the reason why women are said to find computers difficult. They do not correspond to the perceived disciplines needed for today's computer work, and so lead inexorably to the opening statements.

But are the opening statements correct? It is true that it is very difficult to find women in senior positions in conventional computing – only ten per cent of the British Computer Society's members are female. This applies across the board: in manufacturing, software, applications and data processing departments. Is this an inherent feature of being a woman, or the way computing operates? Given the way today's computers respond to users it would appear to be a combination of the two. The important word here is *today's*. To counter the opening statements we shall make a specific assertion of our own. Virtual Reality will make computers as friendly to women as to men.

But there is an additional factor. All the evidence is that from school onwards boys monopolise computers. They not only get to them first, they play their games for lengthy periods, freezing girls out. And the majority of electronic games are designed by big boys for little boys, and involve war, space, killing, bombs, crashes or competitive races; subjects with which girls appear to be uneasy. And in the adult world men tend to take over computers at the workplace, except for clerical and typing tasks. While there was a point, some ten years ago, when European men would not have been seen dead at a personal computer (PC) keyboard, this did not last. Today, PCs and laptops are virility symbols.

So from early days females get less practice on computers, and understandably display less confidence in their abilities to control them. If this were the only reason for the imbalance, the remedy would be relatively simple. But there is a more fundamental problem. As a general rule girls learn to read intuitively by associating sounds with shapes, while boys tend to recognise symbolic letters and words. But the majority of computer interfaces have always made use of abstract and symbolic commands, shutting out those who prefer their information to come from spatial, movement, and auditory sources. By and large this has excluded more females than males.

If anything, this imbalance is worsening. The Universities Clearing House (UCCA) has reported a fall in the number of female computer entrants from 22 per cent in 1981 to 12.6 per cent in 1989, and the number of American women computer science students and professionals is decreasing similarly. The academic courses themselves must be held partly responsible. All too many computer science programmes concentrate on abstract languages first, and only later bring in functional software questions; yet female students themselves say they prefer to learn in the reverse order. Even the language used by female students tends to be interpreted as less 'scientific' or 'rational' than that used by their male counterparts. And there is the widely reported feeling that women are made to feel uncomfortable, even stupid, by their male colleagues or lecturers. Whether this has a basis in fact, or whether it is imagined, makes little difference to the result; by and large women eschew computer science, and as a result role models for younger women are becoming harder to find.

But as most women do not take computer science courses, you have to ask whether there is anything about computers that tends to put women off. There do not appear to be inherent reasons in the machines themselves, so in great part it must be how they are used. In other words the interface. There are shreds of evidence for this view. The first PC both to move away from arbitrary commands and take a whimsical approach was the Apple Mac. Casual inspection by the authors suggests that while some women claim to love their Mac, they never make this claim about any other PC. The whimsical, comically noisy and intuitive interface obviously appeals.

When Myron Kreuger was taking his VR system around county fairs in the seventies, he encountered two singular facts. While the most enthusiastic of the interactors were young women, far outstripping the young men, the most diffident were women with babies and young children. Kreuger explains this by suggesting that women with babies have their protective and survival instincts paramount, and are loth to try new things. And at that time his VR system was controlled by a person at a master console, the players responding to this person's moves. Because a man was seen as being too powerful a woman had to be put in charge; she was known as Big Mother. And this female acceptance of VR as a comfortable medium is reinforced by demonstrations of the Mandala system where women appear to be less afraid of looking silly, at least while everyone is sober. These few examples suggest that any problem lies outside the computer itself, and is centred on conventional interfaces.

There is really no interface with Virtual Reality; no barrier between the user and the computer-generated world. The user is the interface (or the mouse) using normal everyday speech, gestures and sensations, although perhaps not always in everyday ways. Nevertheless, in VR the barrier is down, symbolism is out. So women should benefit. But will they?

In an April 1991 paper Barbara Joans, Director of the Merritt Museum of Anthropology, divided the people involved in Virtual Reality into cyberspace visionaries and practitioners. Regretfully leaving aside her delightful carica-tures of the members of these groups, the message was stark and simple. Although the two elements usually have conflicting aims and objectives, Joans saw them as joint defenders of a predominantly white, male, middle-class ethic. She saw VR as reinforcing existing power structures, unless actions were taken to give access to 'marginal' groups – and these include women and blacks. She based her view on a Virtual Reality conference, where – from their different perspectives – both visionaries and practitioners proclaimed there would be no problem with equality, but both were vague in the extreme as to how it would be achieved.

It may be that Joans has dismissed the 'non-male' side of Virtual Reality interfaces, or she may believe that the 'poli-tics' of power are incapable of changing without a change of rules. But she is surely right when she claims access and encouragement are vital. It is no use having a computer technology which will appeal more to women than to men if women do not know about it. And in Britain at least, neither men nor women have been given anything like the full picture of real Virtual Reality.

British reporting of VR has been dismal. It has been pre-sented as a masculine world of arcade games, real wars and virtual sex, precisely the features that many women find threatening or downright offensive. The difference between America and the UK sits on the cover of Howard Rheingold's recent book *Virtual Reality*. Although the UK and US edi-tions are identical, its UK publishers have added 'Teledil-donics' (the smartech word for virtual sex) to the blurb, while the US cover concentrates on serious issues. This goes some way to explaining why so many British women opinion-formers have chosen to treat Virtual Reality as irrelevant. But Virtual Reality is a multifaceted technology, and women will be the major target for many of its marketing faces.

Fortunately, the British experience is by no means universal. As a technology Virtual Reality is obviously utterly different from mainstream computing, as is evidenced by the large numbers of women involved in America. Despite the diminishing number of female computer scientists, women are at the very top levels of Virtual Reality, on both the technical and applied sides. From prestigious academic establishments like MIT where Margaret Minsky leads tactile feedback research, to IBM itself where Wendy Kellog is a key researcher, and the market leaders in VR equipment, VPL, where Ann Lasco-Harvill is product design manager, women are holding down key positions. And there are many more. This is important. For the first time women are in a position to influence the design, progress and outcome of a computer-related technology. Moreover it is one of *the* technologies of the future.

This point was reinforced in the 1991 San Francisco Virtual Reality Conference, the largest ever held. It was opened, and partially chaired, by Sandra Kay Helsel, a PhD and editor of two highly technical magazines. The token woman? Not a bit of it. Diana Gagnon commanded the complete attention of a big audience. The *Sun* would have assumed that this was because she was an attractive woman; and in the normal course of conventional computer events she would have been talking about women and computers, or something like that. But she held everyone's attention because she was a practical expert on VR entertainment systems. Also a PhD and senior executive with Interactive Associates, a company for whom she negotiates with the hard men of Hollywood, she told how close Virtual Reality games were to public launch in America. And six other major speakers were women, all at least as well qualified as Helsel and Gagnon, with five of them on the leading edge of this, the sharpest-honed technology of them all. And not one spoke about women – all spoke on their technical areas of expertise.

The female attendees were as impressive as the speakers. One, a corporate lawyer from New York with no show-biz experience, moved to Hollywood to produce Virtual Reality films because she saw a tremendous market and did not intend to be left out of it. After only six months she was putting her first deal together. Ellen Renner, Director of Social Service Research at West Washington University, at the conference to see how VR could be used for training in her field, put her finger on the upside and downside of VR as it affects women. While she could see definite training possibilities, the artistic and creative aspects also appealed. But as she was exploring a Sense8 world, swimming among the fish, and shrinking and growing like Alice, she found the helmet too heavy. It also messed her hair, and she recognised this would act as a disincentive for women users.

Workers in the field are well aware of this problem. They also know that marketing will be a huge moneyspinner. Consumers, shoppers, housewives especially, will be targets, as Matsushita's virtual kitchen demonstrates. The key product areas will be cars, groceries, kitchens, decor, toys, clothes, electric appliances, designer goods, and furniture. In today's circumstances we are talking women. It follows that from a marketing executive's point of view it is essential to get women involved in, and using, Virtual Reality systems. This is one reason why so much work is being funded on alternative visual displays and head-tracking devices. The need to get women to like the technology will speed the introduction of wraparound and huge screens, TV glasses or contact lenses, and retinal imaging. At the same time DataGloves will be reduced to conductive nail varnish and rings, and DataSuits shrunk to body jewellery. These changes herald a total change of computing philosophy.

One way and another we have always forced humans to behave like computers; after all we don't type messages, or push mice to communicate with people next to us; we talk, gesture and look at them. In VR we will be able to point, talk, or look at other people or things in virtual worlds. But this will apply as much to a secretary as to a manager; indeed in a virtual world it may be impossible to distinguish between them. It will mean that secretaries will be as capable of entering business information worlds as their 'superiors'. In the long term, if this doesn't alter the relationship between secretaries and executives nothing will. Secretaries will take even more responsibility. But this does *not* necessarily mean that they will be better off. It may well be in employers' interests to dismiss existing managers, and convert their secretaries into poorly paid quasi-managers.

At present the overwhelming majority of women in the workforce have been put into the least prestigious, subordinate, and worst-paid jobs (over seventy per cent of women in full-time jobs earn less than national average earnings, NES 1991). Many of these jobs involve the use of computer keyboards, but with severely circumscribed access and responsibility. But VR could change all that. *Could* is the key word. As Joans pointed out, women must get cultural access. In other words they will have to seize the time, and seize the technology. However, history does not give cause for much optimism. If previous attempts in the workplace have failed, why should it be different this time?

The answer lies in the fact that the technology itself is unique, and once women realise how much it works in their favour, attitudes will change. With an interface which militates against their specialised faculties, men will lose their confidence initiative; it is even possible they will lose interest in the technology as a working tool. Conversely, from schooldays onwards, women should be shaking off their traditional diffidence and replacing it by an unshakable air of confidence. This time around it will be far harder for men to appropriate the technology to themselves in the workplace, if only because it will feel wrong to them. But theory is one thing. It will not just happen. Things will have to be done.

First there needs to be an awareness campaign about how

VR will affect people. This is very much needed in the UK. In direct contradiction to American experience, the number of women speakers at the London version of the San Francisco conference was one – and she was from Moscow. Groups such as Women into Information Technology should be bringing the 'feminine-friendly' aspects of VR to the public's attention. If women's groups as well as the women's pages of newspapers, women's magazines, TV and radio programmes, and women in trade unions took up the challenge of explaining what was possible, instead of privately dismissing the subject as a male fantasy, much progress will be made in a short time. These sources should urge more schools to get hold of VR, and persuade women both to volunteer for VR training, and offer to help in its development. It is a new technology – it is up for grabs.

These are only two 'baby-steps'. The big one will be to operate the systems in a work environment, getting to know where information is, accessing it, absorbing it, using it, and if necessary *withholding* it. The trick in achieving this is confidence. Confidence that at last a computer technology fits stereotyped female thought patterns, and confidence that they can use it at least as well as the men. And if it is argued that this begs the question of families and social relationships, the answer is that only actions of this kind will change them. Waiting for social relationships to change first, on their own, is political and practical cowardice. This time it is unlikely that men will be able to dislodge women from all information sources; and as we know information is power.

Education is also power, and mathematics is a crucial part of education. Yet, even today, girls are dissuaded from taking the subject to its higher levels. But initial research suggests that VR can overcome this – and that it will be a force for gender equality from the earliest ages. An excellent example is algebra. Algebra is a conceptual subject that has always been taught in a very symbolic way. The number of people who hate maths as a result bears witness to the inappropriateness of this method. It would be overstating the case to say it appealed to boys, but over the years it clearly has appealed to girls even less! As we have seen, algebra can be taught in a virtual world. Students can become physical things (bricks or apples). They will personally become numerators and denominators, and cross to the other side of the equals sign. They will physically be a part of the algebraic operation. While this should appeal to both sexes far more than current teaching methods, girls, being process-oriented, will realise that maths can be for them. From a female point of view the important feature is that VR is about teaching by doing, a process-oriented approach, not the goal-oriented method favoured by boys.

Unless you are using a camera-based VR system you can represent yourself in any way you choose: people are genderless, raceless and classless in a virtual world. Users select their own disguises. To that extent women can decide whether they wish to be known as women, or as VR characters. As Meredith Bricken points out, the first thing we notice when we meet a stranger is whether they are male or female. But this may not be possible in a virtual world. What will people notice? Will male or femaleness shine through disguises in body language, or mannerisms? In any event the ability to change or hide gender is not an unimportant point in the context of female computemancipation, certainly until confidence is built up.

Most observers, VR researchers and even US government scientists agree that the development of Virtual Reality will depend more on the psychology of understanding and on creative thinking than on computer sciences. The top VR practitioners are as much creative artists as scientists, and that makes them rare individuals indeed. But the exploiters of the technology are unlikely to make use of these finer artistic feelings. They will go to where the returns are the greatest, and as H. L. Mencken once pointed out, 'No one ever went broke underestimating the taste of the American public.'

This is where women will be important. Surveys of American computer users show that women tend to lose interest in computers if they cannot use them constructively. Men, on the other hand, will happily play games. This female attribute can be turned to Virtual Reality development. At one level Diana Gagnon has been fighting to have sex scenes removed from a new arcade game ... and has succeeded. Talking about her programmers she says, 'These young men come whooping up to me and say, "I've got this wonderful idea," and then go through different ways to kill, explode or maim people. The more blood the better. I try to get them to calm it down, and then do what is practical and commercially viable.'

Gagnon's point is valuable. If Virtual Reality games replace conventional computer games to a great extent, and women continue to be more involved in VR, will this have a general impact on the violent messages that come out of games? Will we see a different set of competitive arenas? Instead of car smashes, perhaps we might have car assembly races, or provisioning a planet, rather than blowing it up. But on another level pressure will be needed to ensure that Virtual Reality gets used in the areas where the social profit outweighs the probable financial return. Here, both genders must fight, but it may be that women will be better placed to take the initiative. After all, they are now in on the design ground floor.

15

Unifier or Divider?
Morals, Society and Religion

Some events shape society: revolutions, famines, natural disasters and wars among them. Most often the changes carry on over a very long period, and go far beyond the original stimulus; for example the loss of almost a complete generation of men in the 1914–18 war. The fact that British women had to do 'men's jobs' during hostilities went a long way to their getting the vote in 1918. In turn this not only changed the political environment, it changed the workplace on a permanent basis. But some technologies change societies out of all recognition. In more recent times these have included the car, the telephone and television.

The car has had enormous effects. When first introduced as the 'horseless carriage' it was a luxury. An early paper questioned its long-term survival because it doubted whether sufficient numbers of chauffeurs could be trained! Yet it has endured to give mobility to hundreds of millions of people, from Rio to Oslo and from Darwin to Nairobi.

The car built upon the trains' freeing of people from their old geographical limitations, and then extended that freedom by several magnitudes. Once, friends and families had to live close together, now they can live hundreds of miles apart. Not long ago we walked to corner shops; now car-centred, one-stop shopping is vital for working men and women. Schools no longer need be community-based. Employees live far from their workplaces and commute.

Roads have changed the physical countryside and cars the shape of cities. Many towns depend financially on car-borne tourists. New laws and taxes have had to be framed, and upheld. Status, class, sex and self-image are all sublimated into the one machine. The car created new industries, among them petrol, accessories, road-building and repairs, insurance and servicing, and promoted international trade and relations, turning frontier crossings into mere staging-posts. Worldwide many tens of millions of people owe their livelihoods to the car. Remove it and modern industrial states would collapse. But the car has also killed and maimed millions; made much of city life unbearable; polluted the atmosphere with gases and noise; is used for crime in general and smuggling in particular, and is probably the largest single source of stress in industrialised countries, outside work and the family.

This has been a hundred years in the making, and still the changes go on. We continue to amend existing laws and bring in new ones. We continue to lay concrete over the countryside, and change the layout of cities. Some young people want to joy-ride, others just want a car. Yet others want fewer cars, for the environment's sake. The car still has the capacity to change society and raise passions. It is a remarkable product.

How will Virtual Reality match up to it? At the outset there is something very different about VR. The car replaced the horse and carriage, television replaced radio, and antibiotics replaced sulfa-drugs. But what is VR replacing? One could say the computer keyboard, but given the number of wide-ranging uses it will have, this is like claiming the car replaced the horse-drawn hearse, ignoring all the other carriages.

Rather than replacing, it is likely to be an 'adding-to' technology, as was the telephone. It will add to television and the movies. It will add to computers, medicine, robotics, education and training, among other things, and it will increase efficiency and improve public and private services. But the lasting consequences are generally caused by secondary effects. For example television has changed our view of war, our appreciation of entertainment and our use of spare time. What will be the equivalent Virtual Reality effects?

The car led indirectly to the dispersal of families. Virtual Reality may put the process into reverse as we make fewer car journeys. There may be a lot of virtual visiting. Driving the twenty miles across London can be two hours' worth of sheer trauma. Families will be able to get closer without their meeting so often; it will be far easier for them to collide in a virtual space. The same applies to friends. And as we have seen from the effects on work, shopping and schools, VR will replace other car use, as well as public transport use. In turn this will create unemployment within the car, transport, and related industries. A percentage of this will be long term, leading to an increase in the underclass. But the car effect will also save the countryside from more concrete, cut pollution, make cities more attractive and so easier to live in. This may reawaken an interest in the city and, after a lengthy period, start to reverse inner city decline.

Governments may well decide to tax Virtual Reality, partly to increase their tax base, and partly to discourage (or encourage) some of its uses. New laws will almost certainly be needed, others amended, and a new International Virtual Law Enforcement Agency will probably be necessary. If the European Community is still in existence, a debatable proposition, the laws and taxes will be Community-based. Virtual Reality will create jobs in electronic engineering, programming, servicing, graphics and information base work. VR's impact on medicine will be responsible for increased longevity with better health, which in turn will create extra poverty as wealth creation will be unable to keep pace with the increase in the ageing population. Gender will become a problematic area, with VR keeping everyone guessing; it will be a cross-dresser's paradise. The home will become one constant thing in our lives, although it will be the base for at least two realities, day-to-day and virtual.

Although having more than one reality sounds confusing, in fact it is part of everyday life. Take Salman Rushdie's *Satanic Verses*. To a devout Muslim parts of it are an insult to Islam. But Western readers often find it difficult to see what the insults actually are, even when they are pointed out. This is no mere difference of opinion or interpretation. The book exists in two very different realities simultaneously, one Islamic, one not. It is no surprise that a religion is involved. Most religious worlds are grounded in their own realities, and find it difficult to comprehend the realities of others; this explains why ecumenical movements just creak along.

Looked at in the cold light of day, this also means that most of our realities are artificial. Otherwise how could the same real things appear to be so different to different groups of people? Our single real-world reality turns out to be like white light, composed of all the colours of the spectrum. The difference is that world realities are made by people, while light exists, and religions claim this to be a starting point of creation.

Religion is the linchpin of many societies. Its tenets govern relationships between people, tell you what is allowed and what is not, and act as a unifying force against intruders. Many a society's cultural heritage has been based on religion. The similarity between Florence and Istanbul in terms of the abundance of religious art is striking. Cathedrals, mosques and temples the world over are the grandest of buildings containing the greatest of treasures, even when the religion itself claims they should not be like that. Florian Brody, an Austrian VR consultant, suggests that cathedrals were early examples of a Virtual Reality, their size and scale out of all proportion to ordinary life, reinforcing the other-worldly aspects of Christianity. Indeed, much Western music is Church-based and many of its most striking paintings and sculptures are dedicated to religious symbols. It is one way that a religion can demonstrate that it alone has the keys to heaven.

Up until now religion has had the monopoly of other

realities and other worlds, to which priests have acted as guardians as well as interpreting the true word. These other religious worlds cannot be entered in life, they can only be anticipated as a reward for piety. The ability to create new, different worlds from their visions is the skill of artists; but their worlds are part of our reality – we can never actually enter and live in them, even for a short time. In that respect they are like the religious worlds of heaven and hell. Virtual worlds, however, are created with the express purpose of being entered. In religious or spiritual terms this is not something to be dismissed lightly. Most religions have but a single God, responsible for creation. Suddenly people can enter and interact with other people's creations and visions, or make their own visions concrete.

It is clear that the contents of a virtual world must be moderated by the influence of the real one. If you create a virtual seven-legged dog, you are using both the model of the dog, and the model of a leg from the real world. It is thus unlikely that theologians and religious 'politicians' will concern themselves unduly with this aspect; after all it is only an extension of God's world and work. But they may be concerned that there will be alternative worlds into which people may slip to find solace – personally tailored heavens – and that people will be able to play God in their own virtual world, as limited as that world might be. These represent potential challenges to organised religions like no others before them.

So how will organised religion cope with Virtual Reality? The precedents are not encouraging. When scientists such as Galileo sought to explain natural phenomena in a rational form rather than as divine interventions, organised religions saw science as a threat. Even today Darwin's view of evolution is challenged in parts of America because it contradicts the Book of Genesis. Islam has also treated science at a long arm's length. And later, when the theatre, movies and television showed us different worlds, some religions disapproved to the extent of trying to control or even ban them. At the outset, religions will probably view VR in the same way as they viewed Galileo's ideas, by seeing it as a danger, and opposing it vigorously; the work of the devil, and that sort of thing. Not only is a secular heaven a direct challenge to religious monopoly of that place, but the ability to play God – especially if it became habit-forming – is blasphemous.

But when religious leaders realise they can use VR for their own purposes, this attitude will change. VR could well provide the greatest proselytising events since the Crusades. Why have a revivalist meeting when you can get people inside the stories? Why tell people about Mohammed or Christ when they can meet and talk to them, or at least representations of them? If you want to persuade Christians that Islam is evil, or Muslims that Judaism is Satan incarnate, why not set up stories where the user is the hero, becoming involved in the battles, the despair, the treachery, the revenge, and the elation? They will be part of the history, part of the ancient enmities. Nothing could be more persuasive than this sort of intense religious historic immersion. It will be the modern equivalent of a 'vision', but available to all.

It will also become possible to have virtual religious ceremonies in a virtual church, mosque, synagogue or temple. Certainly VR will be able to give the impression of a crowded congregation, not an unimportant consideration, and being interactive a virtual mass, virtual communion and virtual sermon could be held. There could even be virtual confessions. Would virtual pilgrimages be allowed, maybe for disabled people? If this sounds sacrilegious, just stop, think, and ask – why not? Unless they are blind. Of course people may find a virtual service unsatisfactory, but from the evidence of American television evangelistic services, and with the added intensity of Virtual Reality, this would appear most unlikely.

There will be two additional challenges to existing organised religions. It is perfectly possible to be intensely religious, or spiritual, without being part of a formal religious body. In both Britain and America today there are growing numbers of people who are interested in Virtual Reality because of its spiritual aspects, and who are airing and exchanging their ideas in small magazines and on bulletin boards. They are not only new-agers and commune dwellers, there are scientists, professionals and ordinary people who no longer take part in formal religious ceremonies. They believe immersion into special VR worlds will extend their ability to delve into their own souls: they see VR as a medium for inner exploration. In turn they feel this will enable them to understand the meaning of the external world that much better.

The major challenge is somewhat different. In all probability, Virtual Reality will spawn its own new, formal religions. These may be based on reincarnation, on a form of Christianity, Buddhism or other meditative philosophies, or perhaps on something completely new. But as Virtual Reality is what you want to make it, it could also be a forcing ground for a revival of the old Roman and Greek religions, mysticism, or perhaps witchcraft – both white and black. Indeed, a religion based on the concept of Virtual Realities rather than the contents of virtual worlds would be consistent with pagan ideas such as cargo-cults or the drug-induced religious diversions such as the Berserkers or the Assassins. And a technology that has miracles as a stock-in-trade should have no difficulty in providing an appropriate religious selection.

The rise of the fundamentalist wings of most religions suggests a widespread need for something strong or different, and Virtual Reality certainly fits the latter description. And what better time could there be for a new religion than the start of the second millennium? What will make it so attractive is that like Marxism it will offer heaven on earth; but unlike Marxism it will be able to deliver, both literally and metaphorically.

Religions provide a measure of stability in societies where the pace of change may be too fast for peace of mind. Britain, with its 'established' Church, is exhibiting a long-run decline in church attendances, and this could signify that the long unbroken run of secular continuity and tradition makes everyday culture and reality sufficient for many British people's needs. But circumstances change, and what is the prevailing set of ethics and morals one year can soon be reversed. In 1960s America and Britain, communes and togetherness were good, profit and selfishness bad. But by the 1980s greed was good and collective behaviour, indeed the 1960s themselves, were bad. One year pornography is tolerated, the next prosecuted. Twenty years ago children born out of wedlock were bastards, now they are just children – what will we call them in twenty years' time? Will Virtual Reality be treated in the same way? One year good, the next ignored, and the following year public enemy number one? And even mainstream religions adapt; their central religious precepts and teachings may remain intact, but current social values often prevail on the margins. Think of the Church of England and divorce.

Virtual Reality will attract judgment because it is capable of challenging almost any accepted set of moral values. It will also upset the notion of a consensus reality. (The fact such a consensus does not really exist does not invalidate the proposition that everyone pretends it does, just as everyone accepts that money has value.) One cornerstone of this consensus is that people have to earn their rewards. Whether these come in the form of wages, holidays, retirement, heaven, or just enjoyment, something for nothing is thought to be normally indefensible, if not downright criminal or corrupt.

But in a virtual world you can get everything for nothing, if that is what you want. There is freedom to choose. The choice of the world itself, and the rules which govern it. The choice of its environment, of your fellow world members, indeed the fundamental choice of self, are at odds with all the disciplines that societies demand from their members. In Virtual Reality, if you don't like the company, the relationships, the laws or the world itself, you just slip into a new world.

Y. Masuda, the vice-president of OKI Electric Industry has

translated this into everyday experience. 'As Virtual Reality becomes widely available, human beings will be able to acquire experiences with no danger and little effort. The information they acquire will be nearly equivalent to the actual experience. Users may come to confuse the virtual experience with the actual one. This can cause anxiety as the individual loses the ability to judge between the reality and the simulation. One problem has already materialised. If a person is allowed to interact intensely with a machine that submissively responds to their desires, they may become too accustomed to it. Eventually, an individual may lose the ability to communicate with people who don't behave as he wishes, and become incapable of carrying on a normal relationship with other humans.'

In other words Masuda is saying that virtual worlds can provide too much, too easily and that this impairs people's ability to operate with other, 'normal' people. This opens the door on to two moral questions. The first is whether it is a bad thing actually to be able to get what you want, even if it is only virtually? The second is that even if we decide it is a bad thing, do we have the right to prevent people from having it?

Part of the argument revolves around the fact that people are expected to work. Masuda may be more worried that the VR 'victim' cannot communicate with his managers than he is about the man's family relationships. But the work ethic was forged when there was no option; if crops were not planted and harvested, clothes made, or defences maintained, it meant death. But life today is different. It is possible to live without working. However, if everyone took this attitude survival would indeed be impossible. So the worry is whether people will continue to work hard if comfortable VR worlds are available to them. In other words this question is not about the right to work – it is about the right *not* to work.

The moral question goes deeper than this, however. It reaches down to the roots of why we live as we do. Is life supposed to be a struggle, or is this something we have imposed upon ourselves? What would happen if we had everything – would it make us better people? Would the urge to compete and to fight wars disappear? If so, why not have long spells in a benign virtual world? The answer is that on re-entry all the familiar problems will come flooding back. Virtual worlds are as perfect as lovely dreams – and as ephemeral. In any event people *do* live in the real world; *do* have to maintain relationships with the people around them and *do* have to have a measure of self-worth. All of us have to be equipped to face these challenges, and virtual worlds may well remove the incentive, and the means, to cope with the rigours of everyday reality. But there is more, however, to Virtual Reality than hiding. It will be the information technology par excellence.

There are two distinct ways of using Virtual Reality-based information. As it is so accessible, and almost everyone will be capable of both getting and understanding it, all information might be made available to everyone as a free service. It could go directly into homes as well as to public buildings, post offices, schools, hospitals and libraries, which would all have access points. This is probably an unattainable Utopia. It implies some form of socialist, perhaps anarchist approach to government. It implies altruism at the highest levels of commerce and the information industry, and it implies the ability of one country to control its information sources, which almost certainly will be impossible. Nevertheless it is a most attractive option.

The second approach will be commercial. It will look upon information as a product, and sell it at a profit. Both the owners of the information and the owners of the VR networks stand to gain from such transactions. Virtual Reality may well be the technology which kick-starts the information market into the twenty-first century. Information is a strange product. After it has been sold, it can be sold again, and again, and again. And when it has been used it does not wear out, and although it gets out of date quickly, all this does is transfer it from the news to the history section. Commercial history suggests that the price of VR information will be set at the highest level possible. The aim will be to get governments, corporations, academia and the military to pay for it rather than individuals.

It is of course possible to provide some free information, and sell the rest, and this is probably how it will be done in practice. And it is certainly possible to price different bits of information differently, perhaps charging schools and hospitals nominal fees. But this would almost certainly require government-imposed regulations or subsidies. Well-off people will be able to subscribe to some information sources, the lower paid will not. In effect there will be information discrimination, as there is today only more so, and for many people a source of power will have been removed entirely.

If we assume that information is intended to raise the living standards and expectations of its users, it follows that the low-paid and disadvantaged will be comparatively worse off. If they become victims of information starvation even more may drop into what André Gorz has described as the class of people with no class, or the underclass. That they exist now is clear from a stroll down London's Strand, or a walk near Washington's White House. They beg, they sleep rough, and their numbers are growing across the industrialised world. Without some token, something to take their minds off their condition, they could become a destabilising force, a threat to the establishment.

Throughout history societies have faced this problem. Their solutions have differed. The Romans tried 'bread and circuses' with some success. Easy, cheap availability of alcohol or drugs (especially hallucinogens) has been another. Religious fervour of varying descriptions or the build-up of a threatening force, sometimes from outside leading to war, sometimes from the inside leading to persecution (Nazis and Jews) are other well-tried methods. If science fiction is to be believed, the underclass will be confined to the sewers or deserts, or placated by 'circuses', ranging from Huxley's combination of Soma and the Feelies, to the more populist *Sex Olympics, Rollerball* and *Running Man*, with all the variations in between. But they all have one thing in common: they are expensive.

Virtual Reality entertainment will be relatively cheap, easy and popular. Even today the average number of hours of television viewing in both America and Britain is awesome. Much of it is known in the trade as 'wallpaper' – it rolls over you. But VR will be infinitely more intense, more gripping, more addictive. A steady diet of interactive soaps, romances and semi-faked news programmes will be capable of lulling all but the most hyperactive adult into a coma. However, as this is mainly television, albeit with an additional dimension, it may not be sufficient. There will be some very angry people out there.

The temptation to provide extra services for such people may well prove to be irresistible. Not only could people be encouraged to join state-sponsored VR religions, take part in gambling game shows or create their own virtual hidey-holes to shut out the cruel world, they could be provided with a range of experiences designed to siphon off aggression, cruelty, jealousy or revenge. Interactive pornography and sex programs leading to kinky meetings in virtual space will prove compulsive to some. Within the relative harmlessness of a virtual world, others will be able to kill, maim, rob, torture, make war, crash, rape and revolt. It will feel real; indeed it is an arguable point, but it actually may be real. Clearly the idea will be to siphon off all possible dangerous responses, and it is no different in principle from unwinding by smashing crockery or throwing brickbats at a model of the boss. However, the strategy is a gamble. Suppose people become addicted to the violence, they might then like to try it for real.

The alternative view is that the gamble would actually be *not* to provide these services. The argument is that the only alternative is to deploy a very large paramilitary police force, and that anyway VR is powerful enough to satisfy destructive urges, without needing to go from the virtual to the real. But there will be a VR alternative. If people are provided with virtual worlds that are all neat and clean, full of happy families living idyllic lives in lovely homes set in unpolluted streets, would users really want to leave them for a grottier real reality? Who could blame them if they revolted? This all means that vigilance will have to be maintained to spot the first signs of information discrimination. And we will need brave people, prepared to speak out against the odds.

Information discrimination could have several effects; those who cannot afford to gain access to information will be handicapped profoundly. Obviously education will be limited, in turn limiting career, indeed training possibilities. But it also will impinge on many other areas of life. Restricted information on diet, medicines or on safety could have severe health consequences. A lack of information in the job market will add to the chances of staying unemployed. A lack of access to financial information could mean money losses, or make it difficult to start a business. If you have sparse information about housing, environmental conditions or schools, moving becomes difficult. If you cannot afford to purchase a weather forecast even a picnic can be hazardous. Information discrimination leads to wider general inequalities. One thing emerges from all this. Either the rumour factory becomes a staple source of information, or an informal (or black) market will develop.

And this is where crime returns to the scene. Theft of electronic information is a difficult enough concept for the law as it stands at the moment. In America several cases regarding computer 'bulletin boards' are creating some sort of havoc within the legal system. Undoubtedly one or more will have to end up at the Supreme Court. But what happens when a virtual person in virtual space 'steals' some of the information? In actual fact it would have to be copied, or remembered. Unauthorised entry, or virtual trespass, or burglary would be better descriptions.

However, it will be very lucrative business. Whenever people or organisations are desperate for a product, they will pay well, and information will be no exception. And if large sums of money are at stake, organised crime will not be far away. Although hackers will have to front the 'break-ins', the ruthless professionals will set up the deals. In thirty years' time street cred could well centre on cipher breaking rather than dope dealing. It could also become as violent as the drug world, with murders of rival gangs as criminals mark out their virtual territories.

Just as the car demanded laws, so will Virtual Reality. It is fascinating to speculate on what might happen to a disputed contract signed in virtual space. It is bad enough if it happens within a single country, but what happens if the space is created between Britain and America? Whose jurisdiction is it? Are such contracts legal anyway? This has enormous implications for shopping, stock markets, gambling or any activity where money changes hands. Not only theft is involved. Copyright will be difficult to enforce, especially with VRAI systems, and there will always be personal questions. For example, if a virtual body is attacked by a virus, could this constitute virtual poisoning, or attempted murder?

The carrying of alien cultures through VR could also be seen as a virus; in all probability an American disease transmitted by a Japanese carrier. Will Virtual Reality be the modern version of the old Silk Route? This time, instead of syphilis and spices, it will be the path of hamburgers and handguns. This international implication of VR is worrying. We have already seen the unintended effects of cultural transfers. Cargo cults are one, and Hollywood films another. They have not only managed to export cultural messages, they have also carried misinformation; for example all too many Middle Easterners believe that all blonde women wearing mini-skirts are either prostitutes or 'easy'.

Information is rarely neutral or objective. In VR it will have to be collected and collated before being put into an agreeable analogue form, and each bit of the process allows for a subjective input. Some information may be omitted, other bits highlighted and the presentation slanted deliberately so as to appeal to certain sections of the population. But overall, the ethos of the international information bases being run across worldwide VR networks will almost certainly be American. As language will not be a problem, the message will be available to everyone. The future of developing countries might well be encapsulated in a picture of a Sherpa wearing a Rambo T-shirt, drinking a Coke.

But will the information reach everyone? Every government has its own policies, and some may not want their people to be well informed. They may try to stop virtual contacts. However, VR will be an immensely anti-authoritarian tool. Granted it might be difficult to break through the locked doors of coded information sources, but once information is made available, the ease with which it can be accessed makes stringent censorship controls difficult. Although at one time it was possible for the Romanians to number all the photocopiers in their country, it is inconceivable that they could have done the same thing with personal computers. And other than the network, and helmet, that is really all that is needed. If the circumstances were right VR could become the electronic equivalent of the samizdat.

Some governments, especially in the lesser developed countries, will want to allow Virtual Reality in, but will not be able to afford the information. Will information discrimination turn into information famine in Africa and parts of Asia? Will VR intensify world divisions, or will it heal them? Could there be a world information bank, giving loans for the less developed countries to obtain technical information? It all depends on people, on whether we allow information to be priced in this way. And in turn the future of the developed countries themselves depends on a similar set of decisions.

Marx saw the world in terms of the means of production. He wanted workers to seize it. But today he would probably feel an urgent need to rethink and rewrite. Information will divide the poor from the rich. Anyone will be able to operate a lathe or welding machine, but the information needed to design the products, the codes in the machine tools, and the patents for the components will be the expensive bits.

That is why correct decisions on the uses of VR are so crucial. It will be one of *the* information technologies over the next century or so, and as such its control should be a matter of concern to every one of us.

16

Addiction and Individuals

The second most striking thing about Disneyworld is that you cannot imagine anyone ever getting ill, let alone dying, there. It is a perfect world, the world of advertising come through to the other side of the screen. Real life is not like that. From cradle to grave we live in the shadow of pain, sickness and death. We have to cope with broken love affairs, rejection and failure. Even if we fail exams, get fired, go broke or get mugged, we are expected to soldier on. But it can all get too much for some people. Life can appear to be hell on Earth for them. Virtual Reality could be their passport to heaven.

There never has been a shortage of people trying to solve their problems with a quick fix. What they use depends very much on fashion, and this has varied over time, and with the prevailing culture. In nineteenth-century Britain laudanum, a highly addictive opium derivative, was intensely fashionable, and well into this century it was possible to buy both cough medicine and baby's teething powders stuffed full of morphine. It goes without saying that alcohol has been popular from the earliest times; fermentation is so easy. Coca leaves are favoured in South America, betel nuts in India, they drink pulch in Mexico, and tobacco has had its moments.

Narcotics, from their name alone, tell you that they help you to forget the world in drowsiness. We took opium to China, and if we have not taken the white poppy to our hearts, we apply its derivatives to the next best thing, our superficial veins. Cocaine is not a narcotic, it is a stimulant: instead of refusing to see the world, the user sees it through rose-tinted glasses. Amphetamines work in the same way, while the hallucinogens alter reality, sometimes alarmingly. And we have created our own instruments, solvents, and designer drugs like PSP and Ecstasy. Most, but not all of these are addictive, but all are illegal, and this combination ensures they are making vast profits for organised crime.

If life is so nasty and brutish, why doesn't everyone take drugs? The truth is that most of us take refuge in something, but for the most part we do not get addicted. Tobacco is just about acceptable, alcohol tolerated in most non-Islamic countries, and the solace of religion is available almost everywhere. Why then do some people get addicted? Medical statistics tell us that one in nine people in industrial countries will need psychiatric treatment at some time during their lives. But it is widely believed in the psychiatric profession that double that number, or twenty-two per cent of the population, is 'vulnerable'. The fact they do not have treatment is either because they do not seek it, could not get it, or have been lucky enough never to meet with the appropriate 'trigger'. So what makes people vulnerable?

Clearly there are hundreds of reasons, but in a nutshell it is about trauma. It may have been an unhappy incident in childhood, indeed an unhappy childhood; an involuntary separation or an adolescent accident (often a seduction or assault) or perhaps rejection. Vulnerable people have a compulsion continually to repeat trying to master this trauma. They have difficulty forming relationships, and drugs can provide a safe alternative. Sex, especially heterosexual sex, involves intimately relating to others; vulnerable people find this difficult. Many vulnerable people feel insecure, they 'know' people will leave them or reject them, and they crave predictability. But in the real world all people are uncertain, and have a habit of creating surprises. So what provides predictability? Religion is predictable, but whereas most people worship from inside themselves, vulnerable people need a relationship with a leader, whom they wish to 'possess'. From an addict's perspective the point of most drugs is that they are very predictable, they are the same each time – *as are computer games*; the parameters are known, there can be no surprises.

But there is more to addiction. There is a paradox. If addicts need predictability why don't they try to find it in something quiet and boring? It is noticeable that addicts nearly always go for something with an element of danger, which creates a frisson of fear. Their kick comes in trying to master the excitement, in the same way that they try to master the original trauma. In short the vulnerable person who will get addicted will 'choose' something which is predictable, but which has that element of fear or danger, and this can be provided by one (or in extremis some combination of) drugs, alcohol, gambling, religion and computer and arcade games.

Virtual Reality will provide intense, absorbing worlds. Most of us will enter them, use or enjoy them, and return to the real world; but then most of us can have a few drinks, then stop. However, alcoholics cannot stop drinking, and vulnerable people will have the same problem with VR games and worlds; they will not be able to stop playing. Two different groups of people are at risk. There are those who wish to avoid relationships with real people (virtual people are far easier to cope with), especially sexual relationships. Either they will hide in asexual virtual worlds, or indulge in virtual sex. The other group is looking for a predictable world, and will play VR arcade or home games. The hook, as compared with ordinary computer games, is their intensity and realism. Once vulnerable people get into them they will feel even more compelled to continue.

The growing literature on arcade and computer game addiction is matched by the one on gambling addiction. Clearly there is an area of overlap. One stark feature of both is that the average age of addicts is falling. Their addiction comes to light only when they steal to get funds, and get caught – one delusion common to both gambling addicts and occasional punters is that they can win! The UK Forum on Young People and Gambling estimates that six per cent of teenagers playing slot machines get hooked. Arcade owners claim that most games addicts have family problems (the owners are trying to indicate that these youngsters would have gone to the bad anyway) but in fact they are confirming a predisposing cause of vulnerability – an unhappy or traumatic childhood. Gambling Anonymous deals with over two thousand under-sixteens every year; one addict described being hypnotised by the 'flashing lights of hope'. And then there is the 'Dungeons and Dragons' syndrome. According to Robert Wright this is characterised by '... a teenager spending weeks on end immersed in a phantasmagoria, emerging to shoot his maths teacher with his father's hunting rifle, then calmly explaining to the police that he was acting on orders from Zataar, the god of final exams.'

This is not just an amusing aside. Vulnerability, especially when it is expressed in an addiction, may be a yearning after passivity. Some psychoanalysts suggest this comes from a desire to be back at the mother's breast, but of course this was never really a passive time. It may have been comfortable, hence the delusion, but suckling – like the rest of life – requires effort. The theory is that many addicts are looking for the present to recreate the past that they never actually had. But deprive some of them of their passivity and they can become violent, even murderous.

The danger lies in the fact that while ordinary computer games are passive, Virtual Reality games are active. The addict, already deprived of part of the passivity by the active VR game (but still playing because of the predictability) finds even this disturbing when interrupted in mid game. What is worse is that the addict's protective shield, the illusion, or delusion, has been shattered. This interruption will provoke an extremely violent response in some people. But there is another area of concern. One of the better-known symptoms of some psychological illness is an inability to distinguish between reality and fantasy, in other words between 'Dungeons and Dragons' and life. If sufferers from such an illness happen to be defending themselves, or killing others in a VR game, one has to ask, will some of them carry on? Will some keep killing in reality? The answer is that given the right circumstances (which is probably an unexpected and unwelcomed interruption) and the proximity of VR to real reality, some psychotherapists believe a small number may well do so.

LSD is known to give flashbacks, often long after the original 'trip'. The hallucinations or visions created by the blood chemistry changes are not chosen by the user; most often they are surreal versions of previous traumas and experiences. Many vulnerable people will find the more psychedelic versions of Virtual Reality similar in effect to LSD flashbacks. Their past experiences will dictate their response, and this may lead to an extremely horrific and frightening hallucination. The consequences of this have always been unpredictable, with documented cases of suicide and gruesome accidental deaths – walking out of windows trying to fly for example. There is every reason to believe the same results will happen if a VR program triggered off a similar hallucination. Quite what would happen to a person in the not too fanciful circumstances of being on an 'acid' trip and trying a VR psychedelic world at the same time, with no real world 'guide', goodness only knows.

The odd thing is that most people who choose to use Virtual Reality as a drug substitute will not become addicted. In the real world, drug substitutes need not be addictive at all. Former addicts use all sorts of devices to keep themselves 'occupied', from crossword puzzles through repetitive behaviour to cigarettes. Obviously the very vulnerable might succumb; but the question is, to what? It will be electronic LSD, using simulated hallucinations. There will be no fixed pattern to these experiences; they might be boring, lively, meditational, challenging or frightening. This is not the reassuring, repetitive, passive world the addict needs; just as LSD is almost certainly not addictive for most people. So by and large, vulnerable personalities who want no surprises do not use LSD in the real world – and nor will they seek out the electronic substitute. The addictive effects of VR will come from its other uses, mainly games.

Battletech is the tank battle game played in Chicago, where four people in a 'tank' play against another, in a shared virtual world. Beth Marcuse, an American sociologist, has made a study of its players in its first year of operation. She found that it was so difficult to play that it takes an average of fifty-six games to learn how to co-operate with fellow tankees. But 'Masters' will have played an average of two hundred and twenty-eight games; almost, if not actually, a sign of addiction. Interestingly, the profile of tankees is very unlike that of ordinary arcade game players. They are mainly male, only children, have high IQs, have an average of fourteen friends and like going to parties. Many play it because they see it as real!

William Bricken discounts the addiction theory. He points out that drugs involve chemical changes, but to get out of a VR 'trip' all you have to do is remove your helmet. If the world were full of 'normal' people this would be true – Virtual Reality should create no lasting difficulties for them. But Bricken's proposition is valid only if you can guarantee a rational response. By definition 'vulnerable' young people are more likely to display irrational responses. Leaving the obvious cases of drug addiction and solvent abuse to one side, the fact that so many strange religious cults – some of them almost 'imprisoning' youngsters – can continue to attract recruits is witness to an altered rationality among a section of the population. It also says a lot about the emotional problems and needs of some young people. Given other circumstances these are the youngsters who could fall into VR very heavily, either in its religious forms, or the more arcane arcade games.

Most of us have played those computer games which start relatively easily but ratchet up the difficulty in successive stages. They are almost impossible to put down; the hours can roll by until the challenge has been met. And this can take weeks. This weakness affects boys (and men) far more than girls and women. It always pays organised crime to deal in addictions; it is a captive market. So if you were an organised crime executive, and you wanted adolescent boys and young men to pour money into your Virtual Reality games, you would build in various addictive hooks. You would start with subliminal messages, words and symbols. Then there would be staged difficulty. There would be delicious surprises, probably mildly sexy and, towards the end, the game will become extremely hard, but not so severe as to be impossible. It is necessary to create heroes who can beat the machine, as they are an important part of arcade game folklore. Even now there are probably animators working to give 'life' to games designed on these lines. But if organised crime wishes to maximise its income from this area, it will also have to tap into the female side of vulnerability. It may well do this by creating and controlling two or three lucrative VR religions, and selecting very charismatic leaders.

If people with psychological problems are likely to be exploited by Virtual Reality, it also will be used in the opposite direction, by treating various psychological disorders. Work is going on in both Britain and America on phobias. At Leeds University Dr Peter Ward, along with Bob Stone, is arranging for virtual spiders to pop out from behind books, or crawl out of virtual pockets or wherever, in an attempt to get arachnophobes to confront their fear, and get accustomed to it. In Washington and Edinburgh arousal states are being monitored while subjects are exposed to virtual dread objects. But while psychologists have suggested that Virtual Reality may be useful in treating a whole range of phobias, the most contentious area is using VR to comfort or treat anxiety sufferers, depressives or schizophrenics.

Not everyone is enthusiastic about VR's uses in treatments. Lawrence Whalley of Edinburgh University is one who, publicly, has urged caution. His most important objections are the dangers of medical paternalism and the loss of freedom of choice of experience. He points out that, 'For patients seeking to understand, as many do, the purpose of their suffering, VR is as unlikely as hallucinogenic drug use to provide access to a deeper reality in their search for meaning.' Whalley believes that much more work needs to be done in understanding the effects of VR before it is used on either psychologically or physically disabled patients.

While this displays admirable caution, it seems highly likely that by the next century patients will be using virtual worlds as both attempted cures and as 'safe havens'. One of the reasons for this is that so many psychiatrically sick people are very unhappy. This will almost certainly concentrate people's minds on the possibilities of VR rather than chemical therapy. Certainly phobias, whether they be of heights, the outdoors, mice, snakes or noise, appear to be susceptible to VR-based intervention. It will provide a feeling so real that sufferers will recognise the risk, but at the same time will also know there is no danger.

Other disorders, especially those which might respond to practical demonstrations, sex therapy for example, should be able to make use of virtual worlds. This may be of great interest in the civilised treatment, as opposed to punishment, of sex offenders. Aversion therapy, using the tactile simulation effects, could be used to help people stop various compulsions, including smoking and drinking. Autistic people, where the world and its communication channels are a closed book to the sufferer, may respond to the concept of an entirely different world; while anorexics may be able to come to terms with their virtual body (or bodies) more easily than their real one. Torture victim therapists may be able to bring a patient to terms with what happened in a virtual world, as may therapists trying to ease the mental agonies of post-trauma victims. In both cases a gentle re-creation of past events, especially with interactivity, may bring peace of mind.

There are questions which have yet to be answered. Would patients in a coma respond to familiar stimuli in virtual worlds more easily than the real one? What, if anything, would Alzheimer's dementia sufferers see? And would catatonic patients prefer the virtual world and respond positively in it? However, ordinary schizophrenics confronted with an almost real other self, or indeed any other imaginary – yet virtually real – person, may react very badly. Such treatment could conceivably induce an acute phase of schizophrenia rather than alleviate the condition. And depressives whose view of their real selves is so dim, might find that a shadow self could tip them over the brink. However, it must be admitted that in both these latter instances the opposite might be the case, and that miraculous cures could follow VR immersion.

Whether the 'safe havens' will be the virtual equivalent of a chemical strait-jacket or a convalescent home remains to be seen. Human nature being what it is, a combination of the two is likely. For example, if VR were to be made available today to 'street people' who have been ejected from psychiatric hospitals, public opinion as exemplified by the tabloid newspapers would applaud loudly, and this could only be described as a strait-jacket. It will be a real dilemma. One very concerned person may argue that because of its addictive qualities VR should not be tried on very vulnerable people, especially as they may not be able to distinguish between the real and virtual worlds. Another, equally concerned person may point out that the only time these gravely disturbed people are at peace is when they are immersed in a virtually real world, and to deny them that would be cruel.

Dr Whalley takes this a step further. He believes that bringing severely physically disabled people back from a VR world into their very limited world, may also qualify as cruelty. However, his caution on the use of VR for quadriplegics and the like appears both to be overstated and, more to the point, overtaken by events. Clever Virtual Reality interface techniques built on tiny muscle movements have already been used in products such as BioMuse to aid communications, and later perhaps they will guide telepresence robots. VR can be used to give paraplegics and quadriplegics the illusion of movement, of walking, running or floating in a virtual world. A person with no arms will be able to pick up and move virtual objects, do a three-dimensional jigsaw, or assemble a car engine. And assuming we have not found a cure for people with degenerative diseases like MS or Parkinson's, they could be given periods of their life when they have total control. Would there be a re-entry problem as has been suggested? Would people not want to come back to real reality? In ordinary life there is no such thing as a good time, unless it can be compared with not-so-good times. And what makes a good time into a memorable time is the ability to compare it with ordinary, or even terrible times. This very human response suggests that only a few would not want to return, and they would be the most vulnerable, perhaps suicidal people. And for them surely a life spent mainly in VR is preferable to no life at all, or is that also debatable?

But Virtual Reality and help for disabled people will go a lot further than taking day trips into virtual worlds. The twenty-first century promises to be the century of miracles. Virtuconferencing will be a breakthrough in social contact, not to mention employment prospects, for the physically disadvantaged. Indeed, for people with disfiguring facial problems, virtual reality will be a lifeline. The Teletact II, and its successor force feedback gloves, should be able to help those with upper body paralysis or injury. VR-style graphics can be prepared so that the visually impaired can use their available sight to best advantage. And above all there are the dreams raised by separate strands of research by Joe Rosen and Greg Panos in California of feeding back healthy nerve transmissions to muscles in paralysed limbs or other parts.

We all have dreams. Some are of the waking variety, and can better be described as ambitions, even if they are unlikely to be fulfilled. Others are the surrealist sleep invaders, that are forgotten almost before they are finished, and yet others – for example Aboriginal Dream Time – are part of life. But dreams and dreamers can be dangerous; it is said of Hitler that he 'dreamed concretely'. Virtual Reality is the technology of dreams, except for one important difference. Dreams carry you along like a piece of flotsam: you, the dreamer, navigate your own way through virtual worlds. In fact where virtuconferencing is concerned VR makes for shared dreams.

When we watch movies or plays, and identify with one of the characters, or part of the action, essentially we are dreaming. But this will be so much more realistic in a virtual world, where we may choose to be a hero or villain, one of the couple meeting under the clock at Waterloo Station, or the clock itself. And it is in VR that many individuals will attempt to live their own dreams. In this respect it could be described as a fantasy amplifier. Individuals will have a choice. They will be able to create their own worlds, join a group of people to do it, or enter one of the commercial worlds that they have rented on a Dream Disc, or which have been cabled into their home on what might well be called the Dream Channel.

The need to unwind, relax and relate to other people is universal in the industrialised world. More than enough therapy consultants and encounter groups have battened on to people who have found it impossible to do any, or all, of these things. Virtual Reality will provide, where perhaps the groups could not. Workplaces will have relaxation rooms where employees will go after a long day at the office or plant, and be lulled by soporific worlds. Or they may be able to work off their aggression in a violent virtual encounter with their boss, or someone from a rival company. What they will almost certainly not be able to do on employers' facilities is get in touch with other employees, or engage in any form of virtual sex. That will have to wait until they reach home, where their relaxation capsules containing the VR equipment wait for them.

The home has been described as an entertainment centre, so think what it will be like when virtucommuting gets going. People will be spending most of their waking hours in the same environment, surrounded by family and friends doing basically the same things. They will need to escape, think new ideas, meet new people, try new experiences. But not too new; that could be overly disturbing. They need to be new and familiar at one and the same time. The Dream Channel and Dream Discs will be designed to meet the needs of people trapped into these circumstances, as well as those returning from work, or indeed those with no work to go to.

These commercial network programs will be created only after extensive market and psychological research, so that they are targeted accurately. This will merely be building on the considerable research already done by television companies on specific social and consumer groups. Network programs will be far easier to use than home-made ones; after all, not everyone will be a talented virtual world maker. Interactive soaps, quizzes and basic cops and robbers drama will provide the backbone of the service, each of them offering a selection of pre-determined roles to the user. And while it will be nice to interact with tropical beaches, climb mountain ranges or scuba dive off the Maldives, the most popular programs are unlikely to be travel. Nor will they be educational, in the traditional sense. DIY, gardening, cooking and specific things like car repairs will have a devoted usership, though, in the last analysis, they will be like clever interactive TV. But fantasies will have a special flavour of their own; they will be the hallmark of Virtual Reality.

By definition, off-the-peg fantasies will be restrictive. If you have a really way-out dream, perhaps having a deep relationship with a sunflower, a personally made world is the only one for you. But as most psychoanalysts will tell you, people's fantasies are depressingly similar. The surroundings may vary, the intent remains the same. Sport and sex for men, friends, appearance and sex for women. Feel the adulation of the crowds when scoring the winning touchdown, holing that final twenty-foot putt or breasting the tape in the marathon. Watch other men and women watching you as you parade the latest spring collection on your way to entertaining your friends at a cocktail party. Egos will get a boost, and everyone will be happy. Or will they?

Some people may be confused by being another person. Most of us have a strong sense of identity, and a virtual persona could challenge that feeling of uniqueness. Indeed, some people may be fearful enough not to take part at all. And those who lack this clear sense of self may become very confused. Other people will feel manipulated by unseen cybernetic forces, indeed feel cheapened by the whole experience. Yet others may not be able to do what they want; production cost considerations will restrict available choices.

Imagine what it must feel like to be a foot fetishist, and then be able to become a virtual shoe; or an SM practitioner to be a whip or perhaps manacles. Sex will be a popular choice. This area will attract the highest advertising rates, and will be where politicians will insert their own little subliminal vision and sound bites. As far as the Dream Channel or Dream Discs are concerned this could become inter-active pornography; a form of a virtual inflatable doll, virtual vibrator, or virtual rubber suit. It will lose some, but not all, of its effectiveness if tactile simulation is absent. From the producer's point of view it is no more difficult to make than ordinary video or animated film. From a user's point of view it is immeasurably superior. The user can choose to approach scenes from any direction or angle, and if it is an animated product, especially one based on body-mapping, also can be part of the 'action'.

If this material is available 'over the counter' or on the public networks, criminals will ignore it; if it is banned and relegated to the informal economy, pornographers will make their usual killing. Difficulties may arise when the authorities wish to make some virtual perversions unavailable. Censorship will always be an issue, but its prevalence will depend on the cultural climate of the time. If VR is being used for underclass control, it is likely that some sadistic VR uses will have been encouraged. So we may see a growth in double standards where, although officially forbidden, the authorities turn a blind eye: rather like today's treatment of soft-drug possession.

Shared dreams include shared sex; Ted Nelson's 'dildonics'. This is not the stuff of the Dream Channel, rather it is a logical extension of virtuphone or virtuconference territory. It is also the extension of what will be virtual dating schemes, which will replace the old computer dating bureaux. It has to be assumed that VR technology will not hold virtual sex back, although the time taken to get into a skin-tight suit dotted with tactile simulators in the appropriate places will certainly preclude any thought of spontaneous love-making, needing as it will the help of considerable amounts of baby powder or oil. However, it should not be sneered out of court. Shy people will use it. Handicapped people will use it. Fantasists will use it. People with the same taste in fetishes will use it. Lovers who cannot get together will use it, and people who would normally go to prostitutes will use it. For many people AIDS will be the key. If AIDS can neither be prevented nor cured by the time virtual sex comes on stream, it will have a clear health advantage; and despite the fact that it has been described as sex minus everything, it will be used.

There are other advantages to virtual sex. Cross-dressers can do so to their hearts' content, without revealing their sex. Bisexuals can do whatever they wish with other consenting adults, without declaring their position (so to speak) and homosexuals need never 'come out'. One partner can take the form the other desires most; indeed each partner can indulge in their own different fantasy at the same time, by defining their virtual space differently. Socially it could have severe repercussions. Birth rates will drop; virtual adultery will become commonplace; prostitutes will become unemployed. But we could also have a new crime wave.

Virtual rape or assault could never be quite as traumatic as the real thing, but it could be nasty. Just patch through to the right code and number, and off you go. The psychological shock of having your virtual body abused could be considerable. However, given the equipment needed it will be impossible to argue a defence that it was a spur-of-the-moment decision. Or hackers could break into other people's worlds, where they may, or may not be having sex; indeed some have already been writing about it. It will be like someone gatecrashing and eavesdropping in your dream, and could turn into virtuvandalism or virtuvoyeurism.

And what if someone alters your virtual persona without your permission? Or perhaps impersonates you in a business deal or a love affair? And how will anyone find this out until it is too late? People can assume any guise they wish. But if it is detected, would this be a virtual form of assault, or even libel, or just breach of your personal copyright? And as we have noted, letting viruses loose in these worlds could cause havoc. Pictures destruct, partners degrade, environments drop into a void, you reach out to touch something and your hand disappears. It could all be very frightening for a healthy, let alone a vulnerable, person.

Of course it is up to everyone to choose whether to use, or not use VR, at least for recreational purposes. It should be no more difficult for non-vulnerable personalities than giving up smoking, or not watching television. Books, people, cinema, indeed the rest of the world, will still exist. There will certainly be enough things to do. In essence it boils down to what people want.

But no government ever allows us to do what we want, not even in the privacy of our own homes. Laws constrain our lives – without them we would probably have chaos. In Britain we cannot drive on the right, in Europe and America on the left. We cannot even do what we want to do when demonstrably it affects no one else adversely. We cannot persuade people to help us commit euthanasia. We cannot take heroin or cocaine. Indeed we are allowed (just) to smoke cigarettes which can harm non-smokers, but cannot smoke ganja, which appears to have no harmful smoke effects on passive smokers.

Freedom of choice, illusory or otherwise, depends on political decision-making. We cannot, indeed must not, assume that democracy will prevail, and that totalitarianism has fallen for ever along with the Soviet Union. The hardest totalitarian regimes have always been religion-based, not just political, and could easily return in our lengthy time-frame. But whatever the political system, choices always have to be made, and in the future decisions about Virtual Reality will loom ever larger. Ultimately the major question surrounding Virtual Reality is who will exercise control over it – and how?

17

Power, Politics and Economics

'Political power grows out of the barrel of a gun,' wrote Mao in the late 1930s, and he had every reason to know. But he omitted to tell us who designed the gun, who provided the gun, and who told the soldier where to aim it. Because whoever they were, they really held the power. Looked at superficially the Gulf War proved Mao's point: guns still win arguments. But a moment's reflection might suggest this to be an oversimplification. Logistics, heavily based on communications and information, plus a touch of Virtual Reality, won the war in one hundred hours: if it had been just gun barrels, or even missile silos, we might well still be fighting in the suburbs of Baghdad.

Information is power. There can be no apology for repeating the slogan; it will be the guiding rule for at least the next century. Some time in the future, one of Mao's spiritual successors will probably declaim, 'Power grows out of the virtual network.' And if you doubt the basic proposition, consider the first target of all rebel groups. They go straight for the television and radio station; in previous times and technologies it was the telephone exchange. Even baddies in Westerns cut the telegraph wires.

Control of the information bases will bring power; as will control of the networks carrying the information. Control of both will mean almost absolute power. It could be a dictator's dream, but it could also be a ruthless industrialist's dream. If it can be used to centralise power politically, it can be used equally efficiently to create monopolies and cartels. The key to understanding the process is that all decisions have to be taken on the basis of having information. All other things being equal, and it has to be admitted they rarely are, the better the information, the better the decision-making. Conversely, decision-making on the basis of almost no information is known as gambling.

The power comes not only from being able to use the information, but also from the ability to provide only that information which you wish to make available, and to choose to whom it will go. Censorship in political terms, good business in industrial language. It need not be overt. A cartel could price its information out of reach of the competition, or perhaps arrange for it to get misleading or irrelevant data. Indeed, control of the information contained in patents, in production and distribution chains, in technological and market research and demographic data, let alone matters such as mineral, agricultural and hydrocarbon resources, is more of a licence to print money than any one national mint has ever been given. So the control of information is crucial, and badly handled it would lead to the greatest concentration of productive and commercial power ever known.

But this would only happen with the connivance of politicians and governments. However, let us not be fooled. If and when this power is granted, it will not willingly be given up. Governments will find that they are powerless, perhaps for ever. Past experience is not encouraging. Transnational corporations have always exercised leverage upon individual countries. From being able to avoid 'excessive' taxation, to getting political favours or subsidies by threatening to place investment elsewhere, these companies have exploited their strength. And much of this lies in their having, and using, information about technical processes, rival companies and other countries' plans and ambitions, to which no one government could have access. This is a considerable advantage.

Cartels, or individual industrial or commercial monopolies, need not have political ambitions to be working against what most people would define as the public interest. The restriction of information to prevent new competition, or drive out existing competitors cannot be good, unless it is believed that the concept of markets and competition itself is antisocial. Whatever the monopolies affect, be it all frozen foods, all shipping, or all legal services, is irrelevant. The effect on prices, service, customer complaints, and choice will be the same, all working against the interests of consumers. But any cartel which is allowed to control either the information bases themselves, or the networks, will be in an immeasurably stronger position to outflank governments, and exploit consumers and the public alike. The possibility of it controlling both information *and* the networks is almost too dangerous to contemplate. It would be the economic equivalent of one nation holding the world's entire stock of nuclear weapons, as well as all the means of its delivery.

Fibre networks will carry much of this information around the world in multimedia and virtual form. And it is this that makes proper control that much more crucial. Most of today's information is unintelligible to all but the specialists. Science, technology and economics have all followed medicine down the road of specialisation. Not only are ordinary politicians (government ministers and civil servants included) and corporate executives unable to understand much of the data, most of the 'experts' find it difficult to interpret. But Virtual Reality will put the same information into instantly comprehensible form, and it will suddenly be possible to get a very much wider understanding of complex issues. If this information were to enter the public domain in a meaningful way competition would be enhanced, and democracy considerably enriched; but if it is suppressed it could lead to political and economic slavery.

One fascinating, and encouraging, historical fact is that although dictatorships have invariably appeared to hold all the cards, they have had very limited time spans. Their control of the military, police, spies, media and even the means of food and industrial production has counted for nothing. They have fallen. But while most communication technologies have been used by totalitarians to reinforce their authority, the advent of PCs plus EMail and computer bulletin boards has moved us in the opposite direction. Information points have become decentralised, making control almost impossible. In any event, once information is in digital form, it is more difficult to alter or censor! So the problems of Virtual Reality and dictatorships are unlikely to be those encountered in the hundreds of novels written about the triumph of the human spirit over the forces of political darkness. They will be more subtle than this.

Dictatorships will use VR in other ways. They might well try to control their underclass with Dream Discs, unless they need everyone to be at work all the time, a most unlikely occurrence. And a Virtually Real prison might appeal. Totalitarianism is characterised by the need for the state to know that everyone believes in its dogma; it was as true of the Inquisition as of Soviet communism. In the former USSR dissidents were placed in psychiatric hospitals on the perverted logic that disagreeing with the state must be a sign of madness. In reality it was just a nasty form of social control. In the future such a regime could sentence its dissenters to a spell of Virtual Reality instead, immersing them in a solid barrage of propagandising virtual worlds. It could be done from home. To ensure the 'prisoner' was actually connected a simple fingerprint detector on the inside of the glove, or some other unique chemical indicator of presence, would be deployed.

The British legal system imprisons more people than any other country in Europe and, leaving aside the morality, by any standards this is an expensive foible. Many people will see the logic in extending virtual prisons to other offences. It will be far cheaper to run a virtual home-prison service; under house arrest would take on a completely new meaning. It will be little different from the American electronic tagging of prisoners. However, it will satisfy the British predilection for punishment far more. It will be possible to recreate a virtual Dartmoor or Strangeways on the one hand, or a virtual open prison on the other. It could even cope with the rehabilitation aspect by providing virtual training and educational courses. The prisoner would not be expected to be like the 'Man in the Iron Mask', wearing a helmet twenty-four hours every day, only for a fixed portion of each day. In fact such an arrangement could provide a suitable alternative to prison for people with pressing social reasons not to be there, for example women offenders with babies, if a custodial sentence is inevitable. There is also a social control aspect. Instead of a period of community service, a careless driver might be sentenced to twenty hours' virtual driving lessons and a new virtual test; or a drug addict could be sentenced to a virtual drug rehabilitation course.

A journalist wrote recently that Virtual Reality could recreate most things, except torture. Would that he were right. Remember poor Winston Smith and the rats in Room 101, in *1984*. But suppose Smith had been frightened of pythons, hippos, or even steamrollers; the regime would have had a bit of a problem. It would have been so much easier to use a virtual python; after all the fear is irrational anyway. Interrogations would be far simpler, and probably much shorter, if you could wrap the subject in a frightening or unpleasant virtual world at the same time. Indeed, subjects could be softened up beforehand by a series of them. It is inconceivable that the uglier regimes will ignore this potential. All that can be said in its favour is that at present it will not give physical pain, although that may not survive the advent of tactile simulation.

Brainwashing is making people believe what the regime believes, or needs its people to believe. This is not just a matter of locking someone up, it is also the steady drip of ideas which get into the marrow of the soul. Virtual Reality will be an excellent mass persuader. Imagine what Goebbels would have done with VR, instead of just film. He would have commissioned worlds where body-mapped actors did all the dreadful things that Jews were supposed to do, with the users involved as innocent bystanders, or as family of the victims. The intensity of the experience would have been awesome.

With VR use people could be persuaded, even more easily than at present, to fight religious wars, to hate and to ostracise. Wartime has always been brainwashing time. People have been persuaded to believe the most bizarre fantasies, like the 'Huns' bayoneting nuns and babies in 1914. But if a government were to invite us to participate in similar virtually real fantasies, every day, in our homes in peacetime, few of us could resist for long; it would not be a question of special vulnerability. Fortunately, many of us live in democracies, and although we must always be careful to distinguish fantasy from fact, the political uses of Virtual Reality will be different in degree.

In a democracy power is supposed to reside in the people. But democratic government was summed up succinctly by Kafka when he wrote: 'Atlas was permitted the opinion that at any time he was free to take the world off his shoulders and walk away. But the opinion was all he was permitted.' The electorate may be an ocean of opinions, but power gets transferred from one outgoing government ship to another incoming government ship. In a democracy, political parties are not primarily in the business of trying to hide the amount of overall information; they are in the business of trying to maximise their vote. This means they try to form our opinions by selecting and interpreting information for us.

Virtual Reality will be used by politicians to persuade voters they have the best policies, and by governments to show the electorate that they are being looked after properly. In this respect, although VR appears no different to television, in practice it will be very much more real. When a politician says, 'Thank you for letting me into your home,' at the end of the Virtual Reality programme, he or she will actually mean it, and unless politicians' habits change radically, they will wish to visit homes regularly, if only to demonstrate they are still alive.

But will it always be a living politician who comes to visit?

This is where the El Cid factor could come into play. El Cid was the Spanish general who rallied his troops to win a battle, even though at the time he was stone dead and tied to his horse. Television has personalised politics; political leaders are judged on their 'performances' at least as much as on their political stance. But all too many politicians are dull, if worthy. In future times when such leaders predominate, it will be possible to enliven their campaigns, raise interest and provoke discussion by taking the body-maps of hero-figures, and using them in virtual political propaganda, scores of years after their death. Among other benefits, this ploy should rally the faithful, an underrated factor in politics. If only body-mapping had been available in their lifetimes, think what seeing, hearing and interacting with De Gaulle, Winston Churchill, Lloyd George, Franklin Roosevelt, Sylvia Pankhurst or Abdul Nasser could do to today's electorates. In the future all it will need is a carefully constructed speech using words the body-mapped leader had used in original political life, an actor 'playing' the body and an avoidance of very close shots. If by some chance the year 2050 has thrown up a group of really boring politicians and religious leaders, our great-grandchildren could well be watching, listening and interacting with Heseltine, Prescott, Delors, Quayle, Pope John Paul II, or their successors.

But there is another benefit for political managers. If a politician were to be taken ill, or was unavailable for other reasons for a key television or VR presentation, a body-map could be pressed into service. This might be especially valuable if chronic illness is involved. With the advent of television it is no longer possible to hide problems such as that which faced the government in the early 1950s when Winston Churchill succumbed to a stroke, and a cover-up had to be instigated. Future cover-ups will have to be behind virtual body-maps.

There will, in all probability, be continuous campaigning. We will be given discs of party rallies where we can choose to be on the platform if we wish, or with the cheerleaders if that appeals more. We will be on the receiving end of political VR games where lovely political aims have to be achieved, and opposition parties are the villains preventing us from winning, and party political VR programs where we interact with the contents. And there will be small overt, or perhaps subliminal inserts into other programs, be they on cable or disc. The whole thing may become very wearing. It may also be treading a fine line between legitimate propaganda and brainwashing. While it is to be hoped that most political parties in democracies would be more subtle than to try to brainwash, attempts to get at schoolchildren through VR games, and adults through repeated VR assimilation, will have to be monitored carefully.

There is an old truism that we get the government we deserve. But this is not really fair; we have to vote without anything like complete information about critical issues, so it is not always our fault. Take economics and defence. In general, the workings of the economy are understood imperfectly, and explained badly, while defence questions remain unput, let alone unanswered, because they are hedged around by security considerations. Virtual Reality will make difficult theoretical concepts very clear. Suppose we were able to get a VR representation of the national economy, using an analogy of, perhaps, a virtual car engine. It would be up to date, operated in real time or as near as you could get to it, and changes in, for example, the money supply, exchange rates, investment or output and all the other factors could be seen, heard and felt. You could ask for predictions, or change parameters to see what would happen by wandering around inside a virtual gasket or gearbox treasury department. Now suppose this were to be made available widely. Would the Chancellor be able to claim inflation was falling when patently it was not? Would an opposition politician be able to tell you the economy was floundering, when clearly it was recovering? It would be much more difficult. Honesty would suddenly strike politics like a thunderbolt.

But how does this differ from economic models available to us now? In one respect not at all, in another completely.

The virtual economy would have to be based on an economic model, which no doubt would suffer from the same defects as the ones we have at present. So to that extent we are no further forward. But, and it is a very big but, you will be able to experience what it tells you at first hand. You will be included in the model – you will be able to feel it, change it, stroke it if you will. You will be inside the processes of change, and be able to see and feel for yourself where the economy was heading, and be able to experiment with changes in, say, interest rates. VR models are not for looking at – they are for experiencing.

VR gives the user far more than an added dimension. And the model is more comprehensible because the analogue used – the car engine – is understandable. (There will of course be other analogues for the mechanically illiterate.) So the model will not have to be interpreted by city economists or politicians, both of whom, obviously, have an axe to grind. And most importantly, as more people come to understand the workings of the economy, so the pressure will grow to improve the model itself, the selection and weightings of the parameters. The main reason this has not been done is public ignorance and indifference; once a campaign is mounted changes will soon follow. In the final analysis the benefits of being able to visualise economic interactions will be every bit as important as those which will accrue to physical visualisations.

In the eighteenth century few people read the very few newspapers. They were informed, everyone else was not. As far as technical information is concerned, we have been living in the eighteenth century for the last three hundred years, relying on others to tell us the truth. Now we can find out for ourselves, and that will shake democratic politics to its very foundations. And if you can do it with the economy, then VR modelling of education, health, transport and foreign policies should be a relative doddle. Indeed, a VR model of the environment, with its international perspectives and connections, would be a wonderful aid to informed debate.

Defence options have a different constraint, national security. However, the virtual economy solves at least one part of the problem. The British government has had a habit of hiding embarrassing defence expenditure inside apparently unrelated categories. Like exploring a virtual library the economic model will make use of VRAI systems, and take you beneath the national information to regional and local information bases. Providing the model is input–output based, it will explode this defence expenditure, or any other accounting cover.

It has always seemed strange that the very people who are being defended are not allowed to know how decisions are taken, what real costs are, and what options are available. For example, if we knew the cost of VR cockpit research, would we still want it? Indeed would we argue for a SIMNET-type war-game system if it meant that the chunks of Dartmoor and Dorset used for military exercises would return to civilian use? Without the sort of easily digested information that Virtual Reality could provide we will never be in a position to make such choices. And this is but a sample; overall a considerable amount of VR development will depend on governmental economic and political decision-making.

In general terms Virtual Reality will only be used if an economic case can be made for it. This means profits. Charles Brownstein of the American National Science Foundation suggests that in the near future we shall be using voice actuators to control lights and toasters, not because they are 'state-of-the-art', but because they are cheaper and cut manufacturing costs. VR must also meet a need or demand. Brownstein points out that the 'talking car', which wished us, 'Good morning' was a financial as well as a cultural disaster, and is a salutary example of ignoring the market, and using self-indulgent design. It is clear that profits will be made from VR in the entertainment and marketing sectors. Equally clearly VR is needed in the military sector, and governments will ensure the demand is met. But there are some areas in which VR will be needed, but profits look unlikely. One of these involves disabled people.

All the current research shows that Virtual Reality could make real reality that much less traumatic for many physically disabled people. But although the severity of the disability is proportional to the need for VR, all too often it is inversely proportional to the ability to pay for it. So who will pay for the VR day-trips, or the Bio-Muse-type machines? Government could do so, in the same way that it creates revenue for arms manufacturers by contracting to buy their wares. And there is a further similarity. Both weapons and disabled aids have an element of short production runs and bespoke manufacturing, which adds to overall costs. But there the similarity ends. We know about arms industry lobbies; but who will lobby for the disabled with the same weight? It would be tragic, almost inhuman, if a lack of political will and funding left disabled people trapped in wrecked bodies when they could be both exploring virtual worlds, and earning their living using VR equipment. The tragedy would somehow be compounded if arms expenditure incorporating VR were to keep increasing.

As a start, why not pressure government into restricting defence contracts to companies which agree to produce Virtual Reality products for the disabled? In many instances there is a technological and a personnel overlap, so it would be a practical proposition. The problem then arises for government that if help is given to one group, how will it resist the claims from other deserving groups – teachers, doctors, nurses and carers – all of whom could make excellent cases. In most countries the majority of education is state-funded. Again this means governments footing the bill for VR. Certainly they should be encouraging experimentation, especially among the very young, and people with learning difficulties. But will they do it? The answer is probably not without a great deal of prodding and lobbying, at least to the same extent as in the early 1980s when the same tactics saw small computers put into schools.

However, American and British illiteracy and innumeracy rates are rising. In America almost one-third of children do not graduate from high school, and of those who do almost three-quarters of a million are functionally illiterate. In the new English 1992 reading and writing tests almost a quarter of seven-year-olds were unable to do either. Clearly something is wrong. It may be that people's attention spans have diminished, while at the same time distractions have increased. The American military is changing its training procedures by using more visual and perceptive systems because of this problem. VR may be seen as a saviour by schoolteachers and the military alike.

Health matters come into the same category. The technology will be there and the need will be there; but will the money needed to match the one to the other be there as well? Virtual Reality could well reinforce the two-tier health market. Those who can afford to pay for cancer-cell killing telepresence robotics will survive, those who rely on state or basic insurance health services will not. This will be even more the case regarding preventative medicine and regular check-ups and monitoring. And care of the elderly, of whom there will be far more, will be another area where income and wealth will dictate service.

It could be argued that these inequalities are all too evident today, so what is the problem? The difference is that in the future Virtual Reality applications could be used to close these gaps. They could help make survival of the fittest, rather than the richest, a matter of course. They could ensure that all children get a better chance to achieve something near to their potential. They could ensure a more meaningful life with enhanced dignity for all elderly people, and one with some pleasure and more purpose for all disabled people. But they can also widen each of these gaps by default.

This is where the moral side of politics emerges. If a technology can do things which by common consent are thought good, does government have a duty to promote it? And if it does, does this extend to making it available to all those who need it? If the answer is no, we go on as we are doing in the West at present, allowing our underclasses to build up in mute despair. In the longer run this will be unsustainable without the control measures we have already outlined.

If the answer is yes, the next question is who pays for it? Economics exists to allocate scarce resources, but the decisions of what is allocated, to whom and where, are taken by people. Yet even if it were possible to get millions of people to agree that resources should be spent on Virtual Reality aids for disabled people, this would not be enough. Continuous pressure will have to be applied to government ministers and opinion-formers, and as other deserving groups will be doing the same thing for their share of VR resources, it will become a scramble. The alternative is to rely on charity as at present, or earmark specific taxes.

Information will be a very lucrative commodity. Organisations will grow wealthy and powerful, and individuals rich, from trading in it. One source of government finance could be to levy a special information added tax (IAT). It would be applied in addition to value added tax, and take account of the value added to raw data by the networks, multimedia, and virtual worlds. The resulting tax revenue would be earmarked for 'moral' Virtual Reality uses, defined as those with a higher social than financial return. In other words the IAT revenue would be spent on the disabled, education, caring and health, and politicians and pressure groups would meet annually to allocate the grants. The principle could be extended to domestic uses of VR, especially if it was felt they were becoming antisocial.

Moral considerations do not stop with domestic politics; indeed they underpin the question of aid to developing countries. Clearly there is little or no money profit to be made, except in terms of 'tied' aid – that is, restricting the recipients' use of the monies to the purchase of goods (often arms) from the donor country. So do we in the industrialised north have a moral responsibility for the less developed south? Probably not, except in terms of guilt left over from old imperial days. But surely there is a common humanity which is stirred by pictures of famine and desolation. This is pertinent because, as we have suggested, if the north denies the use of the information bases to the south, VR could widen the existing gaps considerably further. It could, however, also be used to close them. The non-written, sensory-laden virtual world technique will be the ideal communications medium for countries with low literacy rates and dispersed populations. It could be the ideal teaching tool.

Self-sustaining growth could actually come that much nearer by using VR intensively. But the question re-emerges. Who will pay for the training programs? If the less developed countries were able to pay, they would also be able to feed themselves. Commercial banks already are saddled with mounting debts from such countries, some of which are being written off, and will be disinclined to risk even more. It may be idealistic, but what is needed is a new global Educational Development Bank, perhaps funded by a levy on the global or satellite network providers, which will sponsor the production and distribution of VR materials to the south. Until proper cable networks are installed these will have to be disc-based with home reality engines. To be effective they must be designed in consultation with potential users, not the civil servants of their countries. If even a smidgeon of paternalism is detected, the efforts will come to naught.

It is a very small world, and the politics of many sectors and groups inevitably will have to cross national frontiers. For example environmental questions are often international in character. Virtual Reality use fits into this category, only more so. It will throw up a series of interesting, if challenging, multi-governmental points.

Databases are global. Using existing AI systems it is possible to interrogate computer files the world over on the same network. Knowbots can be despatched, and not only report back to base, they can collaborate with each other, and return home with differently enhanced information. Virtual (and multimedia) information bases will be based on these networks. What happens if one country has almost no laws controlling the dissemination of information, but all the others do? How can knowbots be stopped? The answer is that there is no way, short of depriving them of electricity! In this case the country with lax laws will become an information-haven, in the same way we have tax-havens today. Anyone wanting to get information would apply to the VR systems in that country. And if subjects complained, could they get redress?

And what redress does a person have if virtual impersonation or fraud, plagiarism, assault or simple negligence occur during an international virtuphone or virtuconference? The resolution of such incidents will make international air accident law seem quite simple. Ultimately, there will have to be a corpus of codified case law and legislation. Which body will do this is unclear. It is conceivable that the European Community would lay down technical standards for transmissions, and it would almost certainly wish to be involved in the setting of other laws. It might be possible to define an European Virtual Space (EVS) which could be used to simplify legal procedures in simple intra-Community criminal cases such as fraud. However, there is a range of subjects which will require legislation, but where different countries, including EC member states, with different ethical and moral values, will pull in opposite directions. Nevertheless, the EC will probably issue several directives on these matters.

Over the coming years Virtual Reality will demand that a considerable number of hard choices be made, not least in these ethical fields. But looking on the bright side the 2020s may be the decade when moral philosophers suddenly find themselves on the right end of the job market for the first time in almost two hundred years.

18

Choices

Virtual Reality itself is as neutral as the ball in a tennis match. It is what people, individually and collectively, do with it that matters. So what will we do with Virtual Reality? Well, as we say in the title, we can use it either to glimpse heaven, or to envision hell; remembering all the time that the road to hell is paved with good intentions. Virtual Reality will not create a democracy, but if used wisely could help one. VR does not push us into a dictatorship, any more than a barrel of a gun, but it makes it easier for a dictator to take, and then keep, power. Virtual Reality will not create a society hooked hopelessly on arcade games, any more than heroin or cocaine have their barbs into all of us, but many people will be at risk. And whether we use the principles behind VR to improve the lot of the disadvantaged, or create more wealth for those who have advantages already, is something we shall have to choose.

In principle, two sets of choices are available. Those exercised by ourselves, on our own behalf, and those exercised by society, ostensibly on our behalf. These choices will range from the trivial to the fundamental. Taken together they will determine how VR will be used, where it will be used, and why it will be used.

There is one simple choice that most of us can take for ourselves. We can decide whether or not to enter into virtual worlds. We can decide whether we wish to play VR games or, to a point, whether we want to virtucommute. We can choose to learn in a virtual space, have a virtuphone in the garden shed, or watch Dream Disc and Channel programs. It is up to us if we try to find our inner selves in a meditational virtual world, with or without our own guru, or pray with our own religious group. If we wish to hide, we can hide. If we wish to share, we can share. If we wish to substitute virtual experiences for those that might be dangerous, in travel or theme parks, it is our choice. If we prefer the essence of sanitised exhilaration to the real thing, there will be no one to stop us. If we want nothing whatsoever to do with VR there will always be the twenty-first-century equivalent of the fireproof Welsh farmhouse to which we can escape.

But what about the people who cannot choose? Vulnerable people, often young, who could be exploited and used by those selling the virtually real products. Can we stand aside and watch them become addicted, especially as we know they are at risk? Should they be protected? If so, are we clear we are not only protecting them from themselves, but we are constraining the freedoms of all the other people who wish to exercise their own choices? Do we have a right to intervene? If so, how do we do it?

Government is supposed to exercise moral judgments, and make choices like these on behalf of us all. But in great part these choices were actually made long ago, by the governments of our parents, grandparents or even great-grandparents. In other words many of our laws are old, anachronistic and essentially useless. Certainly the law of copyright has yet to catch up effectively with electronic transmissions. Anti-pornography laws cannot really cope with porn on computer bulletin boards, and contract law is struggling with simple things like facsimiles and EMail. Yet there is probably a precedent on the use of blotters with relation to smudges and quill pens hidden somewhere in English common law.

Given the nature of Virtual Reality, one of the first effective choices will be about censorship. Should VR be censored; indeed can VR be censored? Censorship implies that someone knows what is good for you, better than you do yourself. Clare Booth Luce summed up the liberal view when she wrote, 'Censorship, like charity, should begin at home; but, unlike charity, it should end there.' And of course there are two sorts of censorship, political and non-political.

By definition, dictatorships must always exercise political censorship because, as they ban all other political parties, there is no legal possibility of expressing alternative political views. And in such regimes it is likely that individual choice in the non-political sphere will also be controlled. There is no doubt that dictatorships will take a keen interest in VR programs, and try to vet and ban them. But there is a great tradition of clandestine underground communication networks within dictatorships, and people will construct their own worlds, and then pass their discs on to others. This may be the only really safe network. With people being able to work, meet, talk, discuss, dance, make love or whatever in virtual spaces, the average paranoid dictatorial regime will always assume they are full of people plotting against it. In order to find out what is going on, presumably it will have to indulge in virtuphone tapping, in the national interest of course. It may also maintain an army of virtual informers, who will infiltrate virtual groups in different virtual spaces, rather like a Stasi ghost network.

But before we congratulate ourselves on our democratic rights, we should remember that political censorship is not confined to dictatorships. Sinn Fein is banned from making its points on British radio or television, yet it is not a banned political party. And political censorship is not being able to watch a television film on nuclear war because a government has pressured the television watchdog into banning it, for our own good, of course.

Will Virtual Reality be censored? As almost all other media and publicly available methods of communication are restricted, there is no reason to think that VR will be immune, particularly as it will be making such a wide-ranging impact. It is possible that the first element of censorship will be religious, citing blasphemy rather than politics, good taste or sex as the reason. From the outset new VR religions will be watched closely, and almost certainly banned in some countries. Other examples of banning might be VR programs said to insult a Christian or Islamic religion, in countries such as Ireland or Iran. With the close links between religion and state in many countries, this sort of grey area will mean the door is always open to real political censorship.

Will pornography be allowed? Probably not in Britain. Some people get hot under the collar about the right of other people to watch sex presentations, so the British, if not the Americans, Dutch, Danes or Swedes will be deprived of the right to exercise their own choices. But this raises several questions. What happens with crossed virtulines, especially if people are having virtual sex? Can there be a complaint, and if so from whom – and to whom? Will there have to be a Virtual Transmission Standards Commission? If so, who will staff it? And what if two people get together for virtual sex in a space between Amsterdam and Wrexham? If this contravenes the law in Britain, but not Holland, can the space be defined as being Dutch? Will it depend on who places the call? And if the people in Amsterdam advertise the event, in other words they promote a virtuconference sex event, or virtual orgy, would people in Britain be able to participate? And if the law makes it illegal, how will they be discovered? Either the British start virtuphone surveillance, or the law ignores it, and is seen to be an ass. If we take the former route we shall be standing on the edge of a very steep, very slippery slope. If we take the latter, why bother in the first instance? The moral is that we must be careful in framing our laws.

As we delve into it, Virtual Reality will be seen to be most unusual. All our traditional knee-jerk reactions will have to be suspended until the consequences of any action have been worked through. Yet in theory the choices remain ours. Governments may try to stop us taking part in some virtual experiences, but will we be able to stop junk Virtual Reality marketing programs being downloaded into our recorders overnight? Trying to stop today's junk mail is difficult enough. And as a considerable amount of junk is likely to be religious or political, will it be protected on the grounds that we must defend political and religious freedoms, and to do otherwise would be censorship? In other words, will we or the government be doing the choosing?

The time is approaching when we shall have to choose to copyright our own bodies, movements and voices. Once a person has agreed to be body-mapped, that map can be used in a variety of places, with or without his permission. Not all producers will scrupulously give the copyright to the performer. And not only actors Equity have reason to be worried. Body-maps of anyone could be stolen or pirated. Clearly, well-known people, especially film-stars, apparently participating in interactive VR pornographic programs, would make those programs veritable gold-mines.

It is clear that in Virtual Reality truth becomes a precious commodity, and trusting may be dangerous. Even if you believe you know the other people in your virtual space, you have to keep in mind the fact that Virtual Reality is the perfect place to impersonate other people. We can choose to do this with body-maps, or just by adopting a virtual disguise and pretending to be somebody else. This is where the Mandala system using video pictures of users scores over the more creative graphic-based variety. But while most of it will be harmless fun, for example men and women swapping gender under the cover of being aardvarks, it could also be used to defraud, or mislead. (Some people may be committed aardvark fetishists of course, and we may actually see a virtual 'animal lib' claiming 'Virtual aardvarks have rights too'.) And if we choose to combat this whole area by having laws controlling VR impersonation, these will probably restrict the right of everyone to adopt different guises.

But if we can simulate other people, why don't we simulate other things? Virtual Reality is the technology of dreams, so let us dream. Whenever conflicts look likely, why don't we choose to disarm the generals, and fight virtual wars instead? We could go back to the days when champions represented us: the days of David, Lancelot, or Tommy the pin-ball wizard; only nowadays our champion would be Ed the BattleTech master. Virtual war would be one of the greatest benefits in the history of the human race.

There is a point in the film *War Games* where the boy who is the central character starts to get worried and asks, 'Is this a game, or for real?' The computer replies, 'What's the difference?' This encapsulates the idea; we could have virtual body-counts, invasions, even nuclear attacks. Providing both sides agree to abide by the result, and they fix the prize beforehand, it would be a great saving in both life and property. Imagine if the Serbs and Croats had been battling virtually. Families would still be intact and Vukovar would still be a typical middle-European market town instead of a rat-infested pile of rubble.

While this is a dream at present, and the armaments industry will move might and main to ensure it stays that way, there must come a time when ordinary people are allowed to speak for themselves, and control their own destinies. If Virtual Reality is allowed to develop freely along wide-ranging information networks it will go some way to overcome language barriers, defuse misunderstandings, nullify brainwashing, and make this form of utopian reality at least a possibility.

We choose to give the media the power of disseminating information. Note this is not just data. In so doing we trust it. We trust that competition means that every strand of opinion can be faithfully represented. We trust that nothing is omitted and that what we read, see or listen to, is true. And we choose to believe the views that most nearly approximate to our own. We shall have to place the same trust in the corporations which maintain the information bases, whether or not they are put into virtual form. But in addition we shall have to believe that the network controllers and the designers who will convert the information into virtual worlds for network transmission are themselves neutral and objective.

We have little option but to choose to believe this. If we mistrust the VR information bases we shall be forced to try to interpret traditional material which, although possible, will be both time-consuming and difficult. The way around this problem will be to have alternative virtual information worlds, reflecting different attitudes and ethics. Indeed, it is possible to envisage at least four mainstream economic virtual worlds: one monetarist, one Keynesian, one Marxist, and one green. The choice would then be ours, we could pick one, or compare them all.

But even with this choice we must beware falling into the paternalistic trap implicit in designing worlds for other people. Outside the purely entertainment fields (where considerable market research is made anyway) the users of virtual worlds should be consulted about their design and content. Designers should find out whether potential users actually want a virtual world in the first place, what they want to do with it, and then how their aims can best be achieved. This is very important in sensitive areas such as education. Virtual Reality systems are 'doing' systems. From both the Bricken and Clarke trials, students who have been in on the ground floor as it were, and helped design the world in the first instance, are enthusiastic about learning within it. And the beauty of a VR education system is that it can be redesigned continually.

But there are other sensitive areas. Will we be able to prevent monopolies and cartels controlling VR information? Indeed, will we be able to choose the path of information as a free good, especially for the underfunded and underprivileged, as utopian as it may sound? But whether free or expensive, VR training programs for developing countries should be tailored to the users' requirements, not based on what the designer believes these requirements are. It is bad enough for a farmer in Senegal to be patronised by a Senegalese official, but to be culturally patronised by a white Londoner would guarantee the failure of the program. We are capable of patronising employees, gamblers, an audience or even our colleagues, friends and family, merely because we try to impose our view of reality on their virtual reality. That will not do at all.

The restriction of other people's liberty to choose is often the worst form of paternalism, and Virtual Reality raises ethical questions regarding such restrictions. If enabling a quadriplegic to visit Hong Kong virtually is said to be medical paternalism, surely depriving the patient of the experience is equally paternalistic. In other words any decision taken on behalf of anyone else, without consultation, can be labelled paternal. But what if a person is not capable of the 'rational' responses which make consultation meaningful? Is there a time when we can say that VR is suitable or unsuitable? This is particularly important in the treatment and control of some psychological disorders. But in these instances there are existing guidelines. Unilateral medical decisions often have had to be taken on whether chemical or ECT therapy should be used, or whether to confine a patient. All that has to be done is to use the same criteria, amended in the light of Virtual Reality's singular properties.

There is, however, an extra dimension. A person might be so mentally or physically damaged that he/she can only 'exist' in a virtual world; the psychological equivalent of a 'life-support' machine. When, if ever, do we switch it off? And what about vulnerable people? It could be argued that we have no right to stop people becoming addicted, if that is what they choose to do. But then comes the sixty-four-thousand-dollar question: do they realise that they will become addicted? If not, they have not made that choice at all, and could well need protection. The odds are that the protection will be based on existing arcade age restrictions, based on the erroneous assumption that it is being young that leads to addiction, rather than accepting addiction is about vulnerable personalities. And the propaganda telling young people that it's bad for them will only reinforce that feeling of danger so necessary in addiction. So, like arcade control and the anti-drug ad campaigns, it will be ineffective. The logical next step will be to use the example of heroin or cocaine addiction, and make the supply or use of VR home and arcade games illegal.

Once the first banning step has been taken, it is that much easier to take the next one, and the next, and the next … Other judgments will be made. Safety officers might like to see us all use virtual fairground and theme park rides, and close the real things. But will the thrill of the roller-coaster be the same if you know, absolutely know, that it cannot come off the rails? In other words is it the sensation that matters, or that little frisson of genuine terror and uncertainty? The same question will apply to virtual safaris or rock-climbs, even to the treatment of phobias. The answer is important. It will tell us just how far we are prepared to live an illusion. The screams that come out of today's virtual space rides tend to suggest that the illusion wins, hands down.

Judgments will also be made about this. It will be argued that participating in virtual worlds is a 'cop-out', an escape into fairyland, which should be stopped for the escapees' own good. There might even be a move to restrict the leisure uses of VR. However, other people will claim that virtual worlds are an essential recharging agent in a rigorous, harsh world, and that their use should be encouraged. What they are really saying is that one person's vision of hell is another's glimpse of heaven.

And will we stop people who want to be in heaven? Will we say, 'You can't stay in that virtual world because it's unfair. If we can't all do it, no one can.' Or perhaps it is more likely that we shall wheel out the establishment psychologists to say that prolonged immersion in Virtual Reality is bad for your state of mind, and the government's chief medical officer to say it is bad for your health – thereby sending thousands of potential addicts scurrying into it. In short, will we choose to indulge ourselves in our dreams rather than our work? For the first time ever we have the technology which allows us to make our fantasies concrete. But will we allow ourselves to exercise this option, or will we act as self-regulators because we see it as too threatening to the coherence of our modern industrial societies?

As things stand at present the providers and carriers of VR information will have the unfettered right to choose what information they wish to make available – and so indirectly affect all our choices on every subject. Is this right? Should we, indeed could we, try to change this? The question is how? How do we exercise this, and indeed all the other choices? In a democracy we say simplistically that we can vote, but we vote for packages of policies, not single issues. In any event we first have to get the subject on to the political agenda, presumably by lobbying, using the decentralised system of VR to good effect. But we can also choose with our cheque-books, by not buying things; by ignoring unsuitable laws – perhaps censorship is the classic example; by withdrawing our labour; by demonstrating, or in dictatorships by risking our lives. Choice can be a complex, difficult and dangerous commodity.

So we have a series of different views and choices to make on them. Can these views be reconciled? Probably not. Should they be reconciled? That is probably not a good idea. These diversities of opinion and experience, even within one family, one street, town or country make for pluralist, interesting and stimulating societies. And this is something we should all want to encourage and share in.

At the start of this chapter we asked the question, what will we do with Virtual Reality? Now we must ask, what will it do with us? Virtual Reality is a remarkable technology. It could appeal to the imaginations of people the world over, and shared imaginations can do nothing but good. And far from levelling out the differences between peoples and countries, it could emphasise them, but through a set of common experiences rather than different languages. These are differences that reinforce common feelings of humanity despite separate cultures, they are not the differences of ambition and power that lead to war. Alternatively, in the wrong hands, Virtual Reality will be a remarkable control and brainwashing instrument. It could be instrumental in turning people against themselves, and into a regimented army fed on a diet of hatred and envy. VR could actively be used to promote division and prosecute war like no other technology in history.

The operative word is *could*. Nothing happens automatically in Virtual Reality, or anywhere else for that matter. We shall have to work hard to make sure that Virtual Reality is used to allow us to glimpse heaven. If we fail, the visions of hell will loom large in our sights. More deadly wars rather than virtual wars; divisions rather than understanding and hatred instead of friendship. As with all other technologies there is the light side – and the dark side – and it is up to us to decide which prevails. There is so much at stake. The aim must be to avoid our great-grandchildren having to make that impotent wish that, like gunpowder and nuclear weapons before it, we could disinvent Virtual Reality.

Virtual Chapter 18

Has anything happened over the past year that has caused us to change our minds about the future of VR, or the choices we will have to make? The answer has to be a firm, unequivocal, NO. Indeed, if anything, hindsight suggests we understated both the technological possibilities and the potential of VR to create both good and evil.

Fundamentally, little has changed. Agreed, there have been unexpected advances and hiccups. The large increases in computer power and speed with no increase in costs. The need not to wait for fibre-optics before VR can be beamed into the home and the SEGA initiative all suggest a far more rapid spread of VR than we anticipated. And that will bring the choices that much closer. On the other hand, there are the disappointments from some of the visual displays and trackers, and the recognition of the difficulties in the senses of touch and pressure. But the technical fundamentals were in place, and remain in place.

The reasons why we believe things probably will move further and faster than we suggested initially are less to do with the technology itself than the world business and political environment. Over the past year the number of large companies declaring an interest, and prepared to back their judgment with money, has been impressive. It appears that, providing the technology looks as though it can fulfil its potential, future funding is already on tap. The European Commission and Japanese and US governments' recognition that VR is an important computer technique is of equal importance.

Indeed, the US President's Office of Science and Technology has identified several areas of US national interest concerning VR applications. While these include the expected medical, work-force training, and helping the disabled, it is more imaginative than most when it suggests using VR to improve the quality of life for the increasing numbers of elderly people, and for motivational techniques. The Office goes on to state that there would be significant international economic advantages accruing to countries that use VR. They suggest that these competitive areas will include training simulations, telepresence, design, manufacturing and engineering, not to mention those motivational techniques.

Yet as important as these developments may be, they will not themselves drive VR into its next commercial phase. The ignition key is the end of Soviet Empire and the end of the cold war. It has changed military objectives. No longer are we in a rigid 'Mexican-standoff' with a clear-cut enemy. No longer is it all about nuclear threats and predetermined missile deployments. An enemy might be anywhere. Flexibility. Preparedness and speed. Rapid response forces. International teamwork. Local battles. Planning. Groundwork. Information collection and use. Pin-point accuracy. And training ... training ... training. These needs will drive VR. As it is, US military expenditure on VR and other simulations outstrips the rest of the world's VR costs put together – and this gap will grow. Providing the new-generation hardware and techniques filter into the civilian sector, and under the new US rules they should, we will see a dramatic spurt in VR use.

And where will it go? Almost certainly into the entertainment sector, and if SEGA and its rivals have their way this will be for home consumption, not just in theme parks and arcades. Not far behind will be the less salubrious pornography, and possibly sex markets. Behind that will be marketing, religion and advertising, followed by manufacturing and robotics. Education, training, health and disabled person's needs will, if the market place is left to its own devices, trail well behind. They will have to wait for the price of the military technology to be driven down by the entertainment industry. In terms of global involvement this suggests an American lead, a Japanese take-over and hopefully a European exploitation of the socially useful aspects, especially on the software side.

These developments are accelerating, and with them the time available to make our choices is shortening. Before too long we shall be overtaken by events. Televirtuality is on the doorstep but there has been little discussion of the ethics or moral choices it poses. Certainly there have been many conferences and books, some of which were excellent, and discussed the philosophy of VR. There was even an American conference on the legal implications which concentrated on copyright and patents. But public law, public protection and some notion of 'the public good' have remained untouched, although the US President's Office of Science and Technology acknowledges that these problems do exist, will get worse, and will need to be confronted. The past year has shown us that far from being disinvented Virtual Reality is in the process of being reinvented. If we want to make the best use of its remarkable properties then we need to look at where it is going – and, more to the point, where we want it to go.

Appendix

This appendix is neither an exhaustive list of products, components and their specifications, nor a list of recommendations. Rather, it is intended to do two things. We hope the sample provides a flavour of the sort of equipment available for VR today, and it is also set out on a 'mix and match' basis. Pick one item from each section and you will end up with a VR system – providing they are compatible.

1 PLATFORMS AND GRAPHICS HARDWARE

a) IBM or IBM-compatible 486/66 PCs, with Local Bus or graphics accelerator boards. Types of these are VESA Local Bus Board; TIGA Highlight Board; SPEA FIRE Board.
b) Sun SparcStation XGL.
c) Silicon Graphics Machines from Elan to Reality Engine II and Onyx.
d) Apple Mac Quadra 900.
e) IBM RS 6000 Unix VRS.
f) Amiga 2000 or more powerful.
g) Expiality. Works with Virtuality and V-PC.
h) Supervision and Provision. Transputer-based i860.
i) Pixel Planes 6 and other massively parallel computers.

2 SOFTWARE (Run-Time/visualisation and design)

a) Sense8 WorldToolKit Version 2. Run-time and design. Runs on 1a, 1b, 1c.

b) SuperScape VR Toolkit. Design. Runs on 1a with TIGA Board.
c) dVS runs on 1c, 1e, 1h.
d) REND 386. Runs on 1a.
e) VIRTUS Walkthrough. Visualisation. Runs on 1d.
f) IRIS Performer/Inventor. Runs on 1c.
g) ALIAS Upfront. Design. Works on 1a.
h) Vream. Visualisation and Design. Runs on 1a.
i) Photo VR. Non-interactive.
j) Canon Renderware. Design. Runs on 1a, 1b, 1d.
k) Mandala. Runs on 1f.

3 PERIPHERALS

A) DISPLAYS
a) Flight Helmet from Virtual Research. LCD-based.
b) Cyberface 3. HDTV resolution but mono-occular.
c) Kaiser. 1000 line CRT.
d) Polhemus Labs. 1000 line CRT.
e) Boom system. 1000 line CRT.
f) Crystal eyes. LCD spectacles.
g) EyeGen 3. Virtual Research CRT-based.

B) SENSORS
a) Polhemus. IsoTrak, FastTrack. 120 Hz. Electromagnetic.
b) Ascension. Flock of Birds 8' cube range. Electromagnetic.
c) Logitech. Ultrasonic.
d) Shooting Star ADL. Potentiometer.

C) INPUTS
a) DataGlove VPL.
b) Vertex CyberGlove.
c) TeleTact. Feedback glove.
d) Rutgers. Piston based force feedback glove.
e) Other inputs such as SpaceBall, GeoBall, Commander, 6-D Mouse.

Detailed below are two rough and ready guides to systems:

Low End Specification for total imersion – not desktop – would look like: IBM Compatible 486/66 + SPEA Fire Board + WorldToolKit + Polhemus FastTrak + Flight Helmet + 6-D Mouse. At today's prices this will cost roughly £30,000, but is falling as we write: tomorrow's prices will be nearer £24,000.

A commercial or corporate high-end system would typically consist of: Silicon Graphics Onyx-based Reality Engine II + WorldToolKit + Polhemus Labs HMD + Polhemus FastTrak + Vertex CyberGlove + Teletact Commander. Again, the cost of around £250,000 is dropping rapidly.

Index

281